Lipoprotein Metabolism

Progress in Biochemical Pharmacology

Vol. 15

Series Editor
R. Paoletti, Milan

S. Karger · Basel · München · Paris · London · New York · Sydney

Lipoprotein Metabolism

Volume Editor
S. Eisenberg, Jerusalem

46 figures and 18 tables, 1979

S. Karger · Basel · München · Paris · London · New York · Sydney

Progress in Biochemical Pharmacology

Vol. 13: Atherogenesis. Morphology, Metabolism and Function of the Arterial Wall. Eds: Sinzinger, H.; Auerswald, W. (Vienna); Jellinek, H. (Budapest), and Feigl, W. (Vienna). X + 352 p., 143 fig., 29 tab., 1977
ISBN 3-8055-2761-5

Vol. 14: Ecological Perspectives on Carcinogens and Cancer Control. Selected Papers of the International Conference, Cremona 1976. Eds: Stock, C.C. (New York, N.Y.); Santamaria, L. (Pavia); Mariani, P.L. (Cremona), and Gorini, S. (Milan). VI + 170 p., 51 fig., 42 tab., 1978
ISBN 3-8055-2684-9

National Library of Medicine Cataloging in Publication
  Lipoprotein metabolism
  Volume editor, S. Eisenberg. – Basel, New York, Karger, 1979.
  (Progress in biochemical pharmacology; v. 15)
  1. Lipoproteins – metabolism I. Eisenberg, S., ed. II. Series
  W1 PR666H v. 15/QU 85 L766
  ISBN 3-8055-2985-6

All rights reserved.
No part of this publication may be translated into other languages, reproduced or utilized in any form or by any means, electronic or mechanical, including photocopying, recording, microcopying, or by any information storage and retrieval system, without permission in writing from the publisher.

© Copyright 1979 by S. Karger AG, 4011 Basel (Switzerland), Arnold-Böcklin-Strasse 25
Printed in Switzerland by Graphische Betriebe Coop Schweiz, Basel
ISBN 3-8055-2985-6

Contents

Introductory Remarks 1

Enzymes Affecting Lipoprotein Metabolism

Augustin, J. and *Greten, H.* (Heidelberg): Hepatic Triglyceride Lipase in Tissue and
  Plasma 5
Glomset, J.A. (Seattle, Wash.): Lecithin: Cholesterol Acyltransferase. An Exercise in
  Comparative Biology 41

Intravascular Metabolism of Lipoproteins

Berman, M. (Bethesda, Md.): Kinetic Analysis of Turnover Data 67
Smith, L.C. and *Scow, R.O.* (Houston, Tex./Bethesda, Md.): Chylomicrons.
  Mechanism of Transfer of Lipolytic Products to Cells 109
Eisenberg, S. (Jerusalem): Very-Low-Density Lipoprotein Metabolism 139
Steinberg, D. (La Jolla, Calif.): Origin, Turnover and Fate of Plasma Low-Density
  Lipoprotein 166
Schaefer, E.J. and *Levy, R.I.* (Bethesda, Md.): Composition and Metabolism of High-
  Density Lipoproteins 200

Catabolism of Lipoproteins

Stein, O. and *Stein, Y.* (Jerusalem): Catabolism of Serum Lipoproteins 216

Effect of Drugs on Lipoprotein Metabolism

Carlson, L.A. and *Olsson, A.G.* (Stockholm): Effect of Hypolipidemic Drugs on Serum
  Lipoproteins 238

Subject Index 258

Introductory Remarks

S. Eisenberg

Circulating lipoproteins are lipid-carrying particles of the blood plasma. Operationally, they are subdivided into several species: chylomicrons, very-low-density (VLDL), intermediate-density (IDL), low-density (LDL), and high-density (HDL) lipoproteins. Each of the plasma lipoproteins is a complex particle of finite dimensions and composition. Different lipoproteins, however, differ markedly in size and lipid content. For example, HDL particles are of a diameter of about 80–100 Å; the chylomicrons may attain a diameter of several microns. Hence, the amount of lipid transported in one chylomicron particle exceeds that in one HDL particle by tens of thousands. In contrast, the number of HDL particles circulating in the plasma exceeds that of all other lipoproteins by many folds. Within each lipoprotein, moreover, there exists a spectrum of particles differing in size, weight and composition. Some of these differences reflect changes induced by metabolism, when lipoproteins loose – or gain – lipids and apoproteins. Other differences represent true heterogeneity of lipoprotein fractions isolated from plasma. Whereas the striking differences among lipoproteins are not reflected by the conventional expression of plasma lipid and lipoprotein lipid levels, they are essential for the understanding of lipid transport in lipoproteins. Thus, interactions among lipoproteins, and between lipoproteins and enzymes or cells can be fully understood only when the lipoproteins are regarded as particles.

The number of protein and lipid constituents assembled together in any lipoprotein particles is very large. About 10–12 different proteins (the apolipoproteins) were identified so far in lipoproteins. These include three A proteins (apoA-I, apoA-II and apoA-III), apoB, three C proteins (apoC-I, apoC-II and apoC-III), apoD, several polymorphic forms of apoE, and apoF. Major lipid constituents are triglycerides, phospholipids, unesterified cholesterol and cholesterol ester. Fatty acids, di- and monoglycerides and several different phospholipid species (lecithin, sphingomyelin and small amounts of other glycerophosphatides) are also present. The lipoproteins are not only complex particles, but are also in a dynamic equilibrium among themselves and with blood and tissue cells. Apoproteins and lipids do exchange among lipoproteins. The lipids exchange also between lipoproteins and cells. Lipoproteins, moreover, are donors and recipients of apoproteins and lipids and continuously change their composition and character during metabolism. Some lipoproteins are produced in the plasma, at least in part, and represent dissociation 'units' or breakdown products of the VLDL and chylomicrons. Consequently, change in level, structure, composition or metabolism of any lipoprotein family may affect the levels, structure, composition or metabolism of other lipoproteins.

Circulating lipoproteins are highly ordered particles. It is usually believed that apolar lipids – triglycerides and cholesterol esters – occupy the core of the lipoproteins, and that proteins, phospholipids and unesterified cholesterol constitute the surface ('lipid core' model). Several molecular models of lipoproteins have been recently proposed. All models use the general 'lipid core' concept for the structure of lipoproteins. Yet, newly secreted ('nascent') lipoproteins are different in composition and structure from circulating lipoproteins, and active lipid transport affects predominantly the core compartment of the lipoproteins. In order to conform with the general 'lipid core' model, newly secreted lipoproteins entering the circulation or lipoproteins altered during fat transport, must change and reorganize. This process involves either the core, or the surface (or both) of the lipoproteins. Therefore, structural considerations must be integrated into each of the many events occurring during lipid transport in lipoproteins.

Quantitatively and qualitatively, the main lipid transported in lipoproteins is triglyceride of exogenous (dietary) and endogenous origin. The triglyceride carried in lipoproteins is utilized by many tissues and may be stored in the adipose tissue. Whether lipoproteins play a role in the transport of other lipids – an in particular cholesterol – is uncertain. On theoretical grounds, lipoproteins may be involved with centripetal cholesterol transport, i.e. from tissues to sites of utilization. Failures in transporting lipids result inevitably in severe pathology, as seen in the rare diseases of impaired synthesis of lipoproteins (i.e. abetalipoproteinemia) or in the much more common hyperlipoproteinemia disease states. Lipid transport in lipoproteins is mediated through the activity of several enzyme systems, and through interactions of the particles with tissue cells. Triglyceride transport is dependent on the activity of an acylhydrolase enzyme system, the lipoprotein lipase (LPL). The enzyme(s) was (were) originally described as heparin-releasable hydrolytic activity which decreases the turbidity of triglyceride-rich lipoprotein solutions. Subsequent studies have shown that it is composed of at least two species of enzymes, of hepatic and extrahepatic origin, and contains activities against triglycerides, monoglycerides and glycerophosphatides. The enzymic activity released into the circulation by heparin is therefore heterogenous. A second enzyme system found in plasma and active towards lipoprotein lipids is the lecithin:cholesterol acyltransferase (LCAT). The enzyme transfers an acyl group from lecithin molecules to unesterified cholesterol to form lysolecithin and cholesterol ester molecules. The combined activity of the two enzyme systems, the LPL and LCAT, affects therefore all lipid classes in lipoproteins. Since changes induced in lipoprotein lipids by either enzyme affects also the apoprotein profile of the lipoproteins, these enzymes should be regarded as lipoprotein-metabolizing systems rather than enzymes with activities limited towards their respective substrates.

Another complexity of the lipoprotein system stems from uncertainties concerned the origin of major lipoprotein species and subspecies. Chylomicrons and VLDL are undoubtedly assembled and secreted from tissues, intestine and liver. There is however great doubt whether LDL and HDL are also primary secretory products of cells. It is significant to note that neither LDL nor HDL has been identified as yet inside tissue cells. Recently, it has been established that LDL is formed in the circulation following the delipidation of VLDL. This path may account for all the circulating LDL in the plasma of humans and animals, i.e. monkeys, rats and guinea pigs. Thus, at least one of the major circulating lipoproteins is normally *not* secreted from cells and cannot be regarded as a primary 'lipid transport' vehicle. More recently, the question whether HDL is secreted by cells has also been challenged. The principal HDL apoproteins, apoA-I and apoA-II (in human), are synthesized in intestinal absorptive cells and are secreted with intestinal VLDL and chylomicrons. These apoproteins, as well as phospholipids, unesterified cholesterol and additional apoproteins (apoC and ?apoE) are freed from the triglyceride-rich lipoproteins during triglyceride transport. The possibility therefore that some, or all, of the circulating HDL is derived from surface constituents of chylomicrons and VLDL must be seriously considered. According to this view, HDL may represent the final form of 'surface remnants' generated during the metabolism of triglyceride-rich lipoproteins and is formed in the plasma compartment. Whereas this view is far from being proven, it explains many observations of the relationships among plasma lipoproteins and between the levels of lipoproteins and the levels of the lipoprotein-metabolizing enzymes. It furthermore enables to regard all plasma lipoproteins as constituents of one process, the transport of triglycerides.

An exciting facet of lipid transport in lipoproteins and lipoprotein metabolism has been discovered when quantitative and qualitative approaches for the study of the interaction of lipoproteins with tissue cells were developed. Again, complex processes take place. Partially hydrolyzed triglyceride-rich lipoproteins (so-called 'remnant' particles) interact avidly with liver cells; LDL in contrast, readily interacts with many cell lines (fibroblasts, smooth muscle cells) but not with hepatocytes. HDL may be metabolized differently in different animal species. Some of these interactions result in internalization and degradation of the particles; in others, the lipoproteins may donate or extract lipids from the cells. With either process, the interaction results in profound effects on the cellular lipid-metabolizing apparatus. In turn, the cells change the number of their lipoprotein recognition sites and thus regulate the extent of further interactions with lipoproteins. The nature of the interaction between lipoproteins and cells has been partially elucidated. Specific cell receptors are involved with these interactions (i.e. the LDL receptor). However, to what extent are the receptors specific for a lipoprotein family is yet unknown, as are their physiological role in lipid transport and lipoprotein metabolism.

It is the purpose of this volume to describe the known, and discuss the unknown phenomena occurring during the life span of a lipoprotein particle. This volume begins with a description of the enzymes involved with lipoprotein metabolism. The second part discusses the complex process of fat transport in lipoproteins. The final chapters discuss lipoprotein degradation and the possible mechanisms of action of drugs on lipoproteins and their metabolism. Many distinguished scientists have contributed to the book. I am honored to acknowledge these contributions. It is only by the effort of the authors that a description of fat transport in lipoproteins, as is currently known, could be achieved.

Hepatic Triglyceride Lipase in Tissue and Plasma

J. Augustin and H. Greten

Klinisches Institut für Herzinfarktforschung an der Medizinischen Universitätsklinik, Heidelberg

Introduction

The clearance of circulating lipoprotein triglyceride is mediated by lipolytic activities at or near the surface of the capillary endothelial wall. These enzymes are probably bound to the endothelium since they are not normally measurable in plasma. However, within minutes after the intravenous injection of heparin, lipolysis is markedly stimulated and a group of lipolytic activities can be assayed. These together are referred to as postheparin lipolytic activity (PHLA). The lipase activities of postheparin plasma have been shown to hydrolyze many classes of lipids including triglycerides, di- and monoglycerides, phospholipids and long-chain fatty acyl coenzyme A derivatives [*Greten et al.*, 1969; *Vogel and Zieve*, 1964; *Greten et al.*, 1970; *Ehnholm et al.*, 1975; *Jansen and Hülsman*, 1973, 1974]. Clinical interest in PHLA derives from its potential use in assessing lipoprotein-clearing capacity. PHLA consists of at least two triglyceride lipases with different molecular properties and different sites of origin namely hepatic triglyceride lipase (H-TGL) and extrahepatic triglyceride lipase or lipoprotein lipase (LPL) [*LaRosa et al.*, 1972; *Fielding*, 1972; *Greten et al.*, 1972; *Augustin et al.*, 1978]. Any clinical or pharmacological study reporting decreased PHLA has to account for the presence of these two lipases in postheparin plasma. The full understanding of triglyceride clearance from plasma requires thorough knowledge of the characteristics and availability of lipases *in situ*. At present, however, metabolic and diagnostic studies of these individual lipase activities in the intact subject must depend on the selective measurement after release into plasma by heparin or similar polyanions.

Plasma H-TGL most probably originates from liver plasma membranes. However, at present no data are available as to the particular site of enzyme synthesis. Recently, evidence was provided that a triglyceride lipase is synthesized in the liver and is delivered to the cell surface through a vesicular transport system [Chajek et al., 1977]. Almost all liver cell compartments contain triglyceride lipase activities which have been partially purified and characterized during the last decade. An attempt is made therefore, to review the characteristics of the different lipase preparations from liver cell fractions for comparison with the plasma H-TGL. Clinical disorders associated with lipoprotein abnormalities will then be discussed in the light of altered H-TGL activity and the functional role of this enzyme in lipoprotein metabolism will be evaluated.

Lipase in Liver Mitochondria

Since the mitochondrion is the site of intracellular fatty acid oxidation, it is conceivable that lipases at this location provide substrates for mitochondrial oxidation. Mitochondria obtained by cell fractionation of rat liver homogenates were reported to contain a phospholipase A activity which catalyzes the hydrolysis of the 1-acyl fatty acid, thus giving rise to the 2-acyl-lysoderivatives [Waite and Van Deenen, 1967]. Scherphof and Van Deenen [1965] reported that both ^{32}P-phosphatidylcholine and, at a greater rate, ^{32}P-phosphatidylethanolamine could be hydrolyzed by this enzyme preparation. In accordance with the finding that highly purified pancreatic lipase catalyzes the hydrolysis of phospholipids [De Haas et al., 1965], the mitochondrial enzyme was also found to exhibit lipase activity with sonicates of a triglyceride and phosphatidylethanolamine mixture [Waite and Van Deenen, 1967]. Whereas phospholipid hydrolysis was found to be completely inhibited by EDTA, triglyceride hydrolysis was not affected even at high EDTA levels. In contrast, lipase activity decreased up to 50% by the addition of *p*-chloromercuribenzoate in the assay medium while phospholipase A_2 activity remained unaffected by this inhibitor. These experiments suggested that phospholipids and triglycerides are probably hydrolyzed by separate enzymes.

Two proteins with lysophospholipase activity – lysophospholipase I and II – were isolated to homogeneity from beef liver homogenates [De Jong et al., 1974]. The two enzyme preparations could be further characterized by different molecular weights, different pI and pH optima. Calcium and EDTA had no significant influence on both activities. Both lysophospholipases exhibited general esterolytic propertis. Lysophospholipase I showed a bimodal distribution with highest specific activity in the cytosol and the mitochondrial matrix and was virtually absent in lysosomes and microsomes. Lysophospholipase II appeared to be a membrane-bound enzyme with its highest specific activity in the microsomal fraction [*Van den Bosch and De Jong*, 1975; *De Jong et al.*, 1976]. No data are available on the lipase activity of these preparations against long-chain triglycerides under comparable assay conditions.

Recently, a lipase activity has been purified 77-fold from rat liver mitochondria using gel chromatography of sonically disrupted mitochondrial material [*Claycomb and Kilsheimer*, 1971]. On Sephadex G-50 and G-200 the enzymatic activity eluted with the void volume, indicating a molecular weight greater than 800,000. This might be due to a large lipid-protein complex and no further investigations on enzyme purity or molecular properties were done in this study. Hydrolysis of triglycerides increased as the chain length of the esterified fatty acid decreased. The activity against tributyrin was more than 100 times greater than against tripalmitin. Partial glycerides were hydrolyzed more readily than the corresponding triglycerides with the exception of tripalmitin. This enzyme preparation, however, did not hydrolyze phosphatidylcholine.

The mitochondrial phospholipase A_2 activity has been the subject of several recent publications [*Waite et al.*, 1966; *Nachbaur and Vignais*, 1965; *Waite and Sisson*, 1971; *Nachbaur et al.*, 1972]. This activity has been attributed mainly, but not exclusively, to the outer membrane. In the studies reported so far, different substrates and emulsification procedures have been used. Therefore, comparison of these activities is still difficult and must await further purification of enzyme proteins in order to differentiate between existing enzyme entities of more than one active site on one lipase molecule. In this connection, the recent work of *Okuda and Fujii* [1968], who demonstrated the interconversion of the rat liver esterase into lipase by sonication with lipid, might be of interest.

Lipolytic Activities in Hepatic Microsomes

Monoglyceride lipase activity had been demonstrated in intestinal mucosal microsomes with a possible role in the intracellular completion of fat digestion [*Senior and Isselbacher*, 1963]. A similar enzyme was also found in liver microsomes [*Carter*, 1967]. In comparison with the soluble fraction, the specific activity of this preparation was 10–20 times higher as a monoglyceride lipase and 3–4 times higher as a triglyceride hydrolase. The pH optimum was between 8.6 and 9.0 with monolaurin emulsified in ethanol as substrate. Using substrates emulsified in gum arabic, these results could later be confirmed by others [*Biale et al.*, 1968; *Guder et al.*, 1969]. With whole liver homogenates as the enzyme source, *Biale et al.* [1968] demonstrated substrate hydrolysis in the order monobutyrin >tributyrin >mono->> di- > triolein. This preference for certain substrates was unchanged with isolated homogenate sediments, microsomes or the soluble fraction. The ratio between optimal activity towards monoglycerides to that towards triglycerides varied between different tissues and led the authors to the conclusion that distinct enzyme entities may be responsible for the cleavage of full and partial glycerides in the tissues investigated. Both activities were predominantly located in the microsomal fraction although the possibility that they were derived from other cell compartments could not completely be ruled out. *Guder et al.* [1969] confirmed these results with similar substrate preparations. Since lipolysis of triolein at pH 8.5 yielded a molar ratio of fatty acids:glycerol lower than 3, it was suggested that not all fatty acids were released into the medium, but might have entered other pathways such as transacetylation reaction. However, this possibility remains hypothetical for the rather crude preparation studied.

An esterase has subsequently been purified 254-fold from rat liver microsomes by acetone and ammonium sulfate precipitation and hydroxyapatite column chromatography [*Hayase and Tappel*, 1969]. This enzyme hydrolyzed glycerol 1-monodecanoate at a rapid rate at alkaline pH. Hydrolysis rates for corresponding di- and triglycerides were one third and one hundredth of that of glycerol 1-monodecanoate. All substrates were emulsified in gum arabic. The enzyme exhibited little activity against long-chain fatty acid mono-, di- and triglycerides. Thus, this esterase activity has a broad substrate specificity which overlaps that of monoglyceride lipase. The enzyme preparation was far from being purified to homogeneity, and activity against phospholipids was not studied.

With Triton X 100 as detergent, *Assmann et al.* [1973a] demonstrated considerable hydrolytic activities against long-chain fatty acid tri-, di- and monoglycerides. As only small concentrations of di- and monoglycerides accumulated during the reaction, it was suggested that complete hydrolysis of triglycerides was mediated by this enzyme preparation. The enzyme activity was almost completely inhibited by prior incubation with protamine sulfate (1.5 mg/ml) or with NaCl at concentrations of 0.3–0.63 μM. Higher NaCl concentrations led to a progressive increase of lipase activity.

Several investigators demonstrated phospholipase activity in liver microsomal fractions [*Bjørnstad*, 1966; *Scherphof et al.*, 1966; *Waite and Van Deenen*, 1967; *Nachbaur et al.*, 1972]. As the triglyceride lipase from postheparin plasma also contains activity against phospholipids, the possible identity of lipase and phospholipase activities in liver microsomal preparations should be considered. Recently, it has been shown that highly purified pancreatic lipase is capable of specifically attacking the 1-acyl ester of phospholipids which raises the question of enzymes with both lipase and phospholipase activity [*Sarda et al.*, 1964]. Indeed, enzymatic activity obtained from liver microsomes could be shown to preferentially hydrolyze position 1 of phospholipids [*Waite and Van Deenen*, 1967]. In these studies, some lipase activity was also found to be associated with this cell fraction when sonicates of triolein and phosphatidylethanolamine were incubated together. The positional specificity of a microsomal phospholipase A_1 was confirmed by *Nachbaur et al.* [1972].

Lipase Activity in the Cytosol Fraction

Lipase activity in the soluble supernatant of rat liver homogenates was detected by *Olson and Alaupovic* [1966] at pH 7.0. This fraction contained 10 times more lipase activity than microsomes and mitochondria with a stabilized Ediol emulsion as substrate. As constituents of this emulsion include monoglycerides and phospholipids and as the production of free fatty acids was not correlated with the disappearance of triglycerides, it is not clear whether the measured hydrolytic rate is due to lipase, esterase or phospholipase action.

Alkaline lipolysis was found to be most active in the soluble part of rat liver homogenates by *Biale et al.* [1968] and *Guder et al.* [1969] with 8–10 times faster hydrolysis of short-chain fatty acid triglycerides than triolein. However, because of several kinetic similarities with the microsomal lipase activity, the authors concluded that the activity found in the 100,000 g supernatant of liver homogenates may be released from microsomes into the cytoplasm during the preparation procedure. On the other hand, the possibility that this activity represents a distinct enzyme could not be ruled out [*Mahadevan and Tappel*, 1968]. These results were confirmed by *Assmann et al.* [1973a] who found alkaline lipase activity at pH 8.0 in the cytosol of rat liver cells. The specific activity of this enzyme was far below that found in plasma membranes but as the bulk of cell proteins is present in cytosol, the authors concluded that the majority of alkaline lipase is probably located in this fraction. However, it was not possible to calculate the degree of contamination of the cytosol with lipase of higher specific activity released from other cell compartments. Lipase activity was measured with sonicated triolein, stabilized in Triton X 100, and the enzyme exhibited activity against tri-, di- and monoglycerides. It was 90% inhibited by prior incubation in NaCl at concentrations of 0.3–0.6 M and in protamine sulfate (750 μg/ml).

Mellors et al. [1967] found lipase activity against triolein in the soluble fraction of liver cells with a pH optimum around 5.0. The specific activity of this preparation was only 1.5- to 2-fold higher than in the homogenate, and the same activity was again detected in microsomes. Some phospholipase A_1 and A_2 and mainly lysophospholipase activity was also demonstrated in the soluble fraction of rat liver [*Waite and Van Deenen*, 1967; *Van den Bosch and De Jong*, 1975]. Since some of the activity was always recovered in the particulate fractions, solubilization of the enzyme by the isolation procedure could not be excluded.

Lipolytic Activities in Liver Lysosomes

Lysosomes have been shown to contain the apparatus for complete degradation of nucleic acids, proteins and mucopolysaccharides. The ability of rat liver lysosomes to digest various lipids has also been well documented [*Fowler and De Duve*, 1969]. Lysosomes are able to hydrolyze extensively phosphatidylcholine, phosphatidylethanolamine, phosphatidylserine, phosphatidylinositol, lysophosphatidylcholine, lysophosphatidylethanolamine, phosphatidic acid, cardiolipin, tripalmitin, 1.2- and 1.3-dipalmitin and monopalmitin when incubated with these substrates in 0.1 M acetate buffer, pH 4.3–4.6. Most substrates were hydrolyzed even in the absence of a detergent with the exception of tripalmitin which was deacylated only in the presence of 5% Triton X 100 and cardiolipin which required 0.1% Triton X 100 for hydrolysis. On the other hand, the detergent inhibited the hydrolysis of phosphatidylethanolamine and monopalmitin. Lysosomes were also found to dephosphorylate phosphatidic acid, α-glycerophosphate and, although rather slowly, the phosphomonoesters of choline, ethanolamine and serine. These data demonstrate that lysosomes possess the ability to digest complex natural materials. After the intravenous injection of nonhemolyzing Triton WR-1339, liver lysosomes are loaded with the detergent and can be removed from the heavier mitochondria by density gradient centrifugation of liver homogenates [*De Duve et al.*, 1955]. During dialysis of these particles against buffer, the membranes rupture and release Triton. Centrifugation then separates lysosomal membranes from the solubilized material. This established procedure yielding membrane-supported and solubilized enzymes, however, does not rule out solubilization of originally membrane-bound activities by the detergent. Differential characteristics of the two fractions of rat liver lysosomes were first reported by *Stoffel and Greten* [1967]. They found cholesterylester hydrolase and lipase activity mainly in the supernatant whereas phospholipase A activity was almost equally distributed in both the soluble and the particulate fraction. Similar results have then been obtained by several investigators who in general demonstrated the presence of membrane-bound phospholipase and lipase activity which was most active at pH 4.4 [*Mellors and Tappel*, 1967; *Mellors et al.*, 1967; *Guder et al.*, 1969]. This particular lipase preparation was not affected by either protamine sulfate, heparin or NaCl [*Mahadevan and Tappel*, 1968; *Assmann et al.*, 1973a]. In the absence of Triton X 100, the enzyme preparation showed considerable esterolytic activity on monopalmitin, was slightly active on diglycerides with a preference for the 1.2-isomer and had no detectable activity on tripalmitin. In the presence of the detergent (5%), monopalmitin was poorly hydrolyzed whereas considerable amounts of free fatty acids were released from di- and triglycerides with a specificity for the α-position [*Fowler and De Duve*, 1969]. When lysosomes were dispersed in Triton X 100 before assaying, lipase activity was partially inhibited, and the pH optimum of the enzyme was shifted from 5.2 to 4.2 as demonstrated by *Hayase and Tappel* [1970]. These investigators also showed that the esterolytic activity largely depended on the pH of the assay. Osmotic rupture of rat liver lysosomes led to solubilization of phospholipase A_1 and A_2 activity which could be partially separated by gel chromatography [*Stoffel and Trabert*, 1969; *Franson et al.*, 1971]. These preparations, however, did not hydrolyze triolein.

Some of these conflicting data may be explained by the finding that acid lipase is activated by acid phospholipids [*Kariya and Kaplan*, 1973]. Activation even occurred in the presence of heat-inactivated microsomes using tripalmitin-Triton X 100 dispersions as substrates. The degree of stimulation would then depend on the Triton X 100 concentration in the medium. Heparin and sulfate ions inhibited the activity, but this was partially prevented by phosphatidylserine. The hydrolysis of tripalmitin, dispersed in Triton X 100, taurocholic acid, gum arabic or lysolecithin was activated to a different extent by phosphatidylserine. These results demonstrate that the interpretation of quantitative aspects of lysosomal lipase activity in crude homogenates and partially purified cell organelles is very difficult in the studies reported so far. Several of the data may be rationalized in the light of differential distribution of activities, detergents and endogenous activators.

Rat liver lysosomal lipase has been purified 1,200-fold from crude homogenates [*Teng and Kaplan*, 1974]. The activity was tightly associated with lysosomal membranes as indicated by its resistance to extraction by aqueous salt solutions. Complete solubilization occurred in the presence of 2% Triton X100, pH 6.0. Purification was accomplished on the basis of enzyme-acidic phospholipid interaction. After binding to carboxymethyl cellulose (CM 52), the activity was eluted with brain phospholipid extracts. The enzyme could be absorbed to a second CM 52 column and subsequently was eluted with a NaCl gradient. SDS disc gel electrophoresis revealed one major stainable band with an apparent molecular weight of approximately 58,000. The pH optimum of the activity was 4.0. In acetate buffer with cardiolipin as activator it was shifted to 4.5. In the absence of added phospholipids, this lipase activity was less than 1% of that in the presence of optimal concentrations of phosphatidylserine or cardiolipin. The enzyme also catalyzed the hydrolysis of p-nitrophenylesters and cholesteryl 3-palmitate, the latter, however, only at about 7% of the hydrolysis rate of tripalmitin and only in the presence of phospholipids. No enzymatic hydrolysis of phosphatidylethanolamine or phosphatidylserine was detected in the presence of Triton X 100. The purified lipase was completely inhibited by heparin. The almost total dependence of the lipase activity on added phospholipids indicates an absolute requirement for these activators. No information is available on the mechanism of this activation. Typical allosteric activation with an increase in the apparent hydrolysis rate constant or alterations of either the substrate surface or the interacting enzyme surface would be possible. The data suggest that acid lysosomal lipase is a true lipase with some cholesterylester hydrolase activity but no phospholipase A activity. Thus phospholipid hydrolysis in rat liver lysosomes, if existent, must be provided by different enzyme entities.

Liver lysosomal lipase is supposed to be deficient in cholesteryl ester storage disease (Wolman's disease) [*Patrick and Lake*, 1969; *Sloan and Fredrickson*, 1972]. As noted in other human genetic disorders, the enzyme deficiency in the extracts of mutant cells is not complete [*Beaudet et al.*, 1974]. With fibroblasts from these patients as enzyme source, the hydrolysis rate of cholesteryl esters from plasma low-density lipoproteins was shown to be one third of that of normal cells [*Goldstein et al.*, 1975]. The low-density lipoprotein high-affinity cell surface receptor sites were not reduced. Defective lysosomal hydrolysis of the cholesteryl ester components of low-density lipoproteins thus leads to the accumulation of unhydrolyzed exogenous cholesteryl esters within mutant cell cultures. As a consequence, suppression of 3-hydroxy-3-methylglutaryl-CoA reductase activity and activation of endogenous cholesteryl ester formation is delayed. These data support the concept that acid lysosomal lipase in cultured human fibroblasts supplies the cells with free exogenous cholesterol, while endogenously synthesized cholesteryl esters are hydrolyzed by different mechanisms.

Lipolytic Enzymes Originating from Liver Plasma Membranes

In the past, considerable interest has been focused on the study of isolated plasma membranes from solid tissues. Most of these studies have involved membranes obtained from liver homogenates. Several isolation procedures have been introduced [*Neville*, 1960; *Wallach and Ullrey*, 1964; *Coleman et al.*, 1967; *Anderson et al.*, 1968] which complicates comparison of conflicting data reported so far.

Studying the uptake of lipids by rat liver cells, *Higgins and Green* [1966] found that unesterified cholesterol, cholesteryl esters and triglycerides of chylomicrons were incorporated by isolated rat liver cells at identical rates. Of all subcellular fractions studied, the plasma membranes showed the greatest capacity per unit dry weight for nonesterified fatty acids and chylomicron triglyceride binding as well as triglyceride hydrolysis. Once free fatty acids entered the hepatic cells, they were apparently metabolized in an indentical fashion, whether taken up from the circulation as such or derived from chylomicron triglycerides. They were recovered in mitochondrial and microsomal fractions. There was almost no accumulation of di- and monoglycerides after hydrolysis of chylomicron triglycerides by liver cells suggesting a rapid pathway in the metabolism of triglycerides to glycerol and free fatty acids at the plasma membrane [*Higgins and Green*, 1967]. Chylomicron triglyceride uptake was significantly enhanced by protamine sulfate. Prolonged incubation of liver cells in saline resulted in a marked release of lipolytic activity, which was activated by heparin. Both activities, the saline-extracted and the native lipase, had a pH optimum of 4.0–4.5 in this study. Of all subfractions of the rat liver parenchymal cells studied, the plasma membrane released the greatest amount of lipolytic activity, which was again stimulated by heparin. A major difference between liver and other tissues is that in the former triglyceride-rich lipoproteins have direct access to the parenchymal cells and do not have to cross an endothelial lining, therefore the plasma membrane-bound lipase may be responsible for triglyceride hydrolysis and subsequently internalization of partially degraded chylomicron remnants. In a rather crude plasma membrane preparation of rat liver, *Guder et al.* [1969] found a heparin-stimulated lipase with maximal activity at pH 7.5.

In 1970, *Torquebiau-Colard et al.* reported that the plasma membrane of rat liver contained a phospholipase A that deacylated both phosphatidylethanolamine and phosphatidylglycerol. This membrane-bound enzyme could be effectively extracted by buffers containing 1 M NaCl. Although the pH optimum of this activity was at 7.5, it was supposed to be identical with a phospholipase A_1 hydrolyzing the 1-acyl esters of phosphatidylethanolamine, most active at pH 9.0 and stimulated by calcium [*Newkirk and Waite*, 1971]. Under the same assay conditions, with no albumin and no detergent present, no lipase, but some phospholipase A_2 or lysophospholipase activity could be detected by these authors. The results were confirmed and extended by the finding that perhaps liver plasma membranes contain a phospholipase A_2 that requires calcium and a pH of 8.0 for optimal activity [*Victoria et al.*, 1971; *Nachbaur et al.*, 1972]. Phospholipase A_2 was relatively heat stable and insensitive to deoxycholate and N-ethylmaleimide. The preferred substrates were in the order phosphatidylethanolamine > phosphatidylglycerol and >>> phosphatidylcholine. No lysophospholipase activity was detected in this study and position 1 of the phospholipids was hydrolyzed only to a minor extent by the membrane preparation. This discrepancy may be explained in part by the different membrane isolation procedures in addition to slightly different assay conditions, although contaminations by other cellular components should also be considered [*Victoria et al.*, 1971].

The phospholipase A_1 was separated from the phospholipase A_2 by gel filtration of ammonia acetone extracts of plasma membranes on Sepharose 4-B [*Newkirk and Waite*, 1973]. The A_2 activity was a large aggregate of lipoproteins and was most active against phosphatidylglycerol. The phospholipase A_1 was smaller in size and preferentially hydrolyzed phosphatidylethanolamine, whereas phosphatidylserine and phosphatidylinositol were cleared by both activities at low rates. It is unexplained why microsomal phospholipase A_1 and A_2 could be distinguished neither from the plasma membrane activities by these kinetic criteria nor by differences in their molecular properties.

Mainly the phospholipase A_1 from the plasma membranes of rat liver was solubilized by heparin [*Waite and Sisson*, 1973a]. 10 µg of heparin/mg of plasma membrane protein were sufficient to solubilize 80–90% of the enzyme activity. The addition of heparin to the assay medium stimulated the phospholipase A_1 of plasma membranes 3-fold whereas the solubilized activity was not further stimulated. The soluble phospholipase fraction from membranes which were not treated with heparin was inhibited by the addition of the mucopolysaccharide to the assay medium. The amount of ^{35}S-heparin bound to the membranes was identical with the concentration required for maximal stimulation and solubilization of the membrane-bound activity. If microsomes replaced the plasma membranes, only 25% of the amount of heparin was attached to these organelles. The data suggest a quantitative relationship between heparin binding and the displacement of the enzyme from the membranes. Protamine sulfate inhibited the heparin-mediated stimulation of the membrane-bound phospholipase A_1 as well as the solubilized activity. The former effect was demonstrated to be the result of binding of heparin by protamine, the latter remained unexplained. Calcium stimulated the activities to variable extents. The enzyme was mainly shown to be a phospholipase A_1, although a considerable hydrolysis of position 2 of phosphatidylethanolamine occurred. No attempt was made in this study to separate the phospholipase A_1 from the A_2 activity. Gel filtration on Sephadex G-200 indicated that the enzyme was heterogeneous in size.

In addition to hydrolysis, the enzyme catalyzed transacylation of fatty acids from position 1 of di- and monoglycerides or phosphatidylethanolamine to an acceptor monoglyceride [*Waite and Sisson*, 1973b]. The addition of Triton X 100 to the assay medium decreased transacylation considerably and also reduced hydrolysis of monoglycerides and phosphatidylethanolamine, whereas the hydrolysis of diglycerides was stimulated. Triglyceride was not formed by this system and also poorly attacked even in the presence of Triton X 100 and albumin. Calcium had variable effects on hydrolysis and transacylation, depending on the lipid composition of the substrate and probably the physical state of the liposomes, which was not clearly defined in the study. Again, the activity eluted as a broad peak from a column of Sephadex G-200. The data were extended by the finding that only the acyl group of position 1 of di- and monoglyceride and phosphatidylethanolamine is recognized by the enzyme as a donor in this system, position 2 was not utilized [*Waite and Sisson*, 1974].

Different results were obtained, however, when the subcellular origin of a lipase released from the rat liver by heparin was determined by other investigators [*Assmann et al.*, 1973a]. In this study, the lipase activity was located primarily on the plasma membrane. The assay system included Triton X 100 and albumin. Maximal activity was detected at pH 9.5 contrary to the data of *Guder et al.* [1969]. In this system, the rat liver plasma membrane lipase exhibited a specific activity of 1.8 U/mg membrane protein. Since there was only a small accumulation of di- and monoglycerides during the hydrolysis of triolein, it appeared that the enzyme preparation contained hydrolytic activity against tri-, di- and monoglycerides. After incubation of the plasma membranes with heparin and subsequent centrifugation, about 85% of the membrane-bound lipase activity was released into the supernatant. In agreement with *Waite and Sisson* [1973a], some activity was solubilized from the membranes without heparin treatment. The lipase released by heparin was up to 80% inhibited by prior incubation in 0.63 M NaCl. At higher NaCl concentrations, there was a slight and progressive increase in activity. Protamine sulfate inhibited both the plasma membrane lipase as well as the heparin-releasable activity. The concentration of protamine required for 80% inhibition was dependent on the amount of heparin in the preincubation medium, therefore heparin probably antagonized the inhibitory effect of protamine.

The observation that bovine milk LPL could be purified by affinity chromatography on heparin, covalently linked to Sepharose-4 B [*Egelrud and Olivecrona*, 1972], was utilized for a further purification of the plasma membrane triglyceride lipase. None of the lipase bound to heparin-Sepharose, which had been earlier released from the membranes by heparin. However, up to 80% of the activity solubilized by incubation of membranes with glycine buffer alone was bound to a heparin-Sepharose column. Elution of the enzyme was achieved by increasing salt concentrations in the eluate. This purification step resulted in a 363-fold increase in specific activity, compared with liver homogenates. SDS or urea which were required for satisfactory disc gel electrophoresis of the enzyme protein, completely destroyed the activity. Therefore, it was impossible to locate the position of the enzyme protein on the polyacrylamide gel.

The existence of lipolytic activity against triglyceride-rich lipoproteins in liver has first been demonstrated by *Korn* [1955]. His criteria for defining LPL activity were based on its activation by heparin, its inhibition by protamine and salt and on the requirement for a serum component, which has later been identified as apolipoprotein C-II [*Havel et al.*, 1970; *LaRosa et al.*, 1970]. The enzyme released by heparin from the plasma membranes of rat liver was not activated by heparin, but was at least partially inhibited by protamine and salt. Most striking, however, was the apparent lack of requirement for apolipoprotein C-II as an activator. The enzyme, therefore, fulfilled some of the criteria, adopted for LPL, which, however, depend greatly on the assay conditions used, but behaved differently in other important aspects. The activity, therefore, seems to be a distinct enzyme entity, differing from the lipase of extrahepatic tissues. Studies on the hydrolysis of phospholipids have not been done with this activity yet, and comparison with the enzyme described by *Waite and Sisson* [1973a] is, therefore, more difficult or even impossible.

Enzymes Released into Liver Perfusates

It has been shown that livers from various animals perfused with heparin-containing buffers release lipase activity into perfusates [*Spitzer and Spitzer*, 1956]. *Boberg et al.* [1964] and *Felts and Mayes* [1967] demonstrated that the liver perfusate obtained in dogs and rats was inconsistently inhibited by high salt concentrations. In subsequent studies, this lipase activity has been characterized as a liver lipase which does not fulfill LPL criteria, i.e. requirement of an apoprotein cofactor, inhibition by NaCl and protamine sulfate. From the experiments of *Condon et al.* [1965] it appeared that dog liver is a relatively efficient site of postheparin lipase release which responds to heparin injection promptly. The activity collected from hepatic veins hydrolyzed chylomicrons and very-low-density lipoproteins to a comparable extent. In addition to this release, there was also a significant removal of PHLA by the liver. Studying the portal vein-hepatic vein difference in lipase activity after heparin injection in dogs, *Whayne et al.* [1969] found a highly effective enzyme inactivation system *in vivo* in the liver. Large doses of heparin blocked the system. These authors concluded that a heparinase is involved in first destroying the mucopolysaccharide heparin with dissociation of the enzyme-heparin complex the lipase then being removed by a second mechanism. With a similar experimental design, using perfused rat liver, this hypothesis was confirmed [*Naito and Felts*, 1970]. In fact, lipase from the perfusate was 3-fold more stimulated by heparin than plasma lipase activity. *LaRosa et al.* [1972] observed that the enzyme released from rat liver by heparin on perfusion responded in a bimodal curve of activity as ionic strength was increased with a maximum of about 1.5 M NaCl in the incubation medium. There was also a slight activation by protamine and pyrophosphate but a progressive decrease in activity in the presence of increasing amounts of high-density lipoproteins in the incubation system. Complete inhibition was obtained with roughly half the high-density lipoprotein concentration of normal human plasma. As also previously shown by *Hamilton* [1964], the preparation was relatively ineffective in catalyzing the hydrolysis of triglycerides of chylomicrons and very-low-density lipoproteins compared to those in artificial emulsions. It was, therefore, concluded that this particular enzyme preparation did not play a functional role in the catabolism of plasma lipoproteins. When the substrate triglyceride concentration was saturating, no inhibition of the rat liver enzyme was found, and it was observed that the relative hydrolytic activity of the lipase against natural substrates was substantial, namely approximately two thirds that of extrahepatic lipolytic activities [*Krauss et al.*, 1973]. The importance of sufficient and comparable assay conditions emerged from these studies in which a substrate for detection of lipolytic activity was used that contained basically 10 mM triglycerides, 1.5% defatted bovine serum albumin, 0.05% Triton X 100, 0.2 M Tris-HCl buffer (pH 8.6) and 0.15 M NaCl [*Krauss et al.*, 1973; *Assmann et al.*, 1973a]. Using this system, the following results were obtained. The lipase released into rat hepatic perfusate by heparin was inhibited up to 80% by prior incubation with 0.6 M NaCl and activity rose steeply to the original value as the NaCl concentration was increased to more than 0.75 M. Variations in the amount of Triton X 100 and albumin in the assay medium caused less satisfactory hydrolysis of triolein. Protamine sulfate in a concentration of 0.75 mg/ml assay media inhibited the activity up to 80% depending on the salt concentration within the substrate. At NaCl concentrations higher than 0.5 M, the activity appeared to be protected from inhibition, and in addition, the protamine sulfate effect was time-dependent. At 27°C, the rate of decline of the activity from the rat liver was only 10% after 10 min, however 90% after 180 min, which indicates that protamine sulfate is acting time-dependently on the hepatic enzyme. Heparin, even in high concentrations had only moderate effects on this inactivation as well as on the salt inhibition of the lipase of the hepatic perfusate. The addition of 50 U of heparin to the hepatic perfusate raised the concentration of protamine in the assay system required for inhibition of the activity from 1.5 to 6 mg/ml. By kinetic criteria, the heparin-released activity in the liver perfusate was identical with the lipase from liver plasma membranes. Therefore, the latter enzyme appears to be the major, if not the entire, source of lipolytic activity released by heparin into rat hepatic perfusate. The enzyme released more than 3 M of free fatty acid per mole of glycerol into the incubation system, indicating a significant accumulation of partial glycerides [*Assmann et al.*, 1973b]. Primary ester bonds were preferentially attacked. A different reaction at positions 1 and 3 could not be established since the enzyme had no stereospecificity for either of the two primary ester bonds in triglycerides. This was confirmed with glyceryldiether monoesters under conditions of substrate saturation. These results, however, were not in agreement with earlier findings that the preheparin plasma of the rat and the heparin perfusate obtained from rat liver exhibited lipolytic activities mainly against monoglycerides and tributyrin substrates and only to a minor extent against tri- and diglycerides [*Fielding*, 1972].

Jansen et al. [1977] confirmed the presence of considerable monoglyceride hydrolase activity in postheparin hepatic perfusates. The activity was completely inhibited by a monospecific antibody raised against the salt-resistant lipase from rat postheparin plasma. It was concluded that monoglyceride hydrolase and triglyceride lipase activities are catalyzed by one enzyme with a higher affinity and a higher velocity with the monoglyceride substrate. The same inhibitory pattern was found for phospholipase A_1 and palmitoyl-CoA hydrolase activities in the perfusate. These data support the concept that H-TGL, released by heparin into rat liver perfusates, may act on a variety of substrates. In a nonrecirculation liver perfusion system, the hepatic releasable enzyme was shown to comprise about 30% of total liver tissue lipase activity [*Chajek et al.*, 1977], although substrate saturation may not have been obtained in this study. The activity was neither inhibited by NaCl nor by protamine but decreased substantially in the presence of serum. Colchicine pretreatment of rats resulted in a marked decline in the rate of enzyme release by heparin whereas the tissue activity slightly increased. This indicated a vesicular transport mechanism from one of the cell compartments to the plasma membrane. Inhibition of protein synthesis in the rat liver by cycloheximide was used to determine the half life of the releasable activity which was calculated to be about 3 h. Pretreatment of the animals with chloroquine to inhibit lysosomal protein degradation was followed by a moderate rise in total tissue enzyme activity suggesting that hepatic lysosomal enzymes participate in the degradation of H-TGL. It should be noted that in dog liver perfusates the presence of an apolipoprotein-C-I-activated LPL has been reported [*Ganesan et al.*, 1976]. In rooster hepatic perfusates, the heparin-releasable triglyceride hydrolase activity consists of at least two distinct lipolytic enzymes, which could be resolved by affinity chromatography on heparin-Sepharose-4 B [*Bensadoun and Koh*, 1977]. One of these activities resembles the salt-resistant H-TGL, the second enzyme fulfills LPL criteria, is stimulated by apolipoprotein C-II, inhibited by 1 M NaCl and by antibodies developed against adipose tissue LPL. In young roosters, LPL is only a minor component in the liver perfusate but accounts for most of the lipolytic activity in mature animals. This pattern may reflect differences in the hormonal states, which is known to influence H-TGL [*Kelley et al.*, 1976].

H-TGL in Postheparin Plasma

Intravenous heparin injection results in a dramatic change in the plasma lipid and lipoprotein spectrum [*Graham et al.*, 1951; *Lindgren et al.*, 1955]. Shortly after heparin injection, the plasma triglyceride concentration decreases for about 2 h, paralleled by a considerable increase in plasma free fatty acids. Within 6–24 h, plasma triglycerides again reach their original values. The pre-β-band in the lipoprotein electrophoresis is diminished, the α-band usually enhanced. Ultracentrifugation reveals that the concentration of triglyceride-rich lipoproteins, chylomicrons ($S_f > 400$) and very-low-density lipoproteins (S_f 20–400) decreases whereas that of low-density lipoproteins (S_f 0–12) and high-density lipoproteins increases [*Graham et al.*, 1951]. At the same time, apolipoprotein C is transferred from chylomicrons and very-low-density lipoproteins to the high-density lipoprotein fraction [*Gustafson et al.*, 1966; *Brown et al.*, 1970; *LaRosa et al.*, 1970].

The importance of this PHLA for the metabolism of triglyceride-rich lipoproteins was demonstrated in patients with familial type I hyperlipoproteinemia who had low lipase activities in their postheparin plasma [*Havel and Gordon*, 1960; *Fredrickson et al.*, 1963] and in extracts from fat biopsies [*Harland et al.*, 1967] assayed under conditions optimal for LPL.

PHLA was believed to be identical with a triglyceride hydrolase originating from several different tissues, LPL [*Korn*, 1955]. The enzyme was activated by incubation of tissue homogenates with heparin and a plasma cofactor which was identified as apolipoprotein C-II [*Lindgren et al.*, 1955; *Havel et al.*, 1970b], inhibited by high ionic strength and protamine sulfate. However, especially these latter inhibitors had variable effects on PHLA [*Fredrickson et al.*, 1963; *Datta and Wiggins*, 1964; *Greten et al.*, 1968]. *Greten et al.* [1969] demonstrated that after heparin injection, monoglyceride and triglyceride hydrolase activities appear and disappear at the same rates. In these studies, monoglyceride hydrolase activity was greater than the latter, less heat-sensitive, unaffected by NaCl, protamine sulfate and normal in type I hyperlipoproteinemia. Hydrolysis of diglycerides was similar to that of monoglycerides [*Greten et al.*, 1970].

In vitro experiments with purified LPL from bovine milk [*Egelrud and Olivecrona*, 1973; *Nilsson-Ehle et al.*, 1973] and rat postheparin plasma [*Fielding*, 1970] established that the enzyme is rather unspecific and shows major activity against tri- and diglyceride substrates, particularly against the α-position acyl groups but retains some activity with 1-monoglycerides. Ester bonds in the α- and β-position of glyceryldiethermonoesters were equally hydrolyzed by PHLA from human plasma [*Greten et al.*, 1970].

In addition, PHLA was shown to be a very active phospholipase A_1 [*Vogel et al.*, 1965] with a preference for phosphatidylethanolamine; the V_{max} was 25 times higher with this substrate than with phosphatidylcholine [*Vogel and Bierman*, 1967]. Deoxycholate and a pH of 9.6 were required for optimal activity. Although after heparin injection the concentration of very-low-density lipoprotein phospholipids declines, no increase in lysophospholipid derivatives could be detected [*Vogel and Bierman*, 1968]. Studies to separate the phospholipase activity from the triglyceride hydrolase in postheparin plasma by means of ultracentrifugation, zone electrophoresis or gel filtration were unsuccessful [*Vogel and Bierman*, 1969]. Postheparin phospholipase activity was reduced by 80% in the hepatectomized rat. In contrast, PHLA was only slightly affected by this treatment [*Zieve and Zieve*, 1972]. Following intravenous heparin injection in humans, the time course of phospholipase activity coincided with triglyceride lipase, diglyceride and monoglyceride hydrolase [*Greten*, 1972]. With double- labeled phosphatidylcholine it could be demonstrated that there was no absolute positional specificity of the activity for either the α- or the β-position in the lecithin molecule. The enzyme was found to be normal in patients with different types of hyperlipoproteinemia. Triton WR1339 totally inhibited the phospholipase activity while NaCl (1 M), taurodeoxycholate and Mg had no such effect.

All these data indicated that in rat as well as in humans PHLA may be heterogeneous and consist of more than one enzyme entity. Comparison of the rat triglyceride lipase of adipose tissue and postheparin liver perfusates and plasma demonstrated that the enzyme from adipose tissue differed in several major respects from the other two activities, including response to inhibition by high ionic strength and protamine and to activation by apolipoprotein C-II in the incubation media [*LaRosa et al.*, 1972; *Krauss et al.*, 1973]. The assay conditions were the same as described in the chapter dealing with liver perfusates. The response of PHLA to increasing salt concentrations during preincubation paralleled that of the postheparin hepatic perfusate. Maximal inhibition was at 0.5–0.6 M NaCl, the activity then rose progressively with higher salt concentrations. The degree of inhibition at all NaCl levels was greater with PHLA than with the enzyme from liver, whereas PHLA from a hepatectomized rat was completely inhibited by NaCl concentrations from 0.5 to 2.0 M. This effect differed markedly from that obtained with plasma from the intact rat and corresponded closely to that observed with LPL from adipose tissue. These observations were confirmed after preincubation of the different activities with protamine sulfate. Adipose tissue LPL and the activity in plasma from the hepatectomized rat were almost completely inhibited by this substance. The enzyme released into the hepatic perfusate was more resistant to protamine and, at NaCl concentrations greater than 0.5 M, appeared to be protected from inhibition. Again, PHLA from the intact rat paralleled that of the hepatic enzyme. These experiments indicated that at least two different lipase activities are present in rat postheparin plasma, one of hepatic, the other of extrahepatic origin. While the extrahepatic lipase was inhibited by protamine and NaCl, the liver enzyme was resistant to these inhibitors under appropriate assay conditions. Different degrees of partial hepatectomy conclusively demonstrated that the protamine-resistant activity in PHLA decreases in direct proportion to the amount of liver removed. This decrease indicated that the liver contributes to PHLA and is the major, and possibly the sole source, of protamine-resistant lipase activity in postheparin plasma. These results then led to the development of an assay system for selective measurement of hepatic and extrahepatic lipase activities in postheparin plasma, which will be evaluated later in this review [*Krauss et al.*, 1973]. *In vitro* experiments with endogenous substrates demonstrated that human very-low-density lipoprotein triglycerides were hydrolyzed by both activities to almost the same extent with a preference for primary ester bonds [*Assmann et al.*, 1973b].

Heparin covalently linked to Sepharose-4 B was found to bind LPL from bovine milk [*Olivecrona et al.*, 1971] under appropriate conditions. Elution of the enzyme was carried out with a linear gradient of NaCl from 0.15 to 1.2 M. The same procedure was applied for human postheparin plasma after removal of very-low-density and low-density lipoproteins [*Greten et al.*, 1972]. Most of the plasma proteins did not bind to the column and were eluted during the application of the sample. After the initiation of the NaCl gradient, a triglyceride hydrolase activity was eluted at an NaCl concentration of 0.5 to 0.75 M. The specific activity increased more than 1,000-fold compared to postheparin plasma. Rechromatography of this enzyme preparation on a second heparin-Sepharose column yielded another 10-fold purification. This purified lipase was maximally active at 0.5–0.75 M NaCl in a triolein-gum arabic system at pH 8.4 and, as an important difference to LPL, addition of preheparin plasma or apolipoprotein C-II did not stimulate the enzyme. Thus the human enzyme exhibited properties of activities isolated from postheparin liver perfusates and plasma membranes in the rat and therefore most likely originated from liver. Further evidence for the hepatic origin of this postheparin plasma lipase was obtained from experiments with hepatectomized dogs [*Greten et al.*, 1974] and from tissue studies performed in humans [*Klose et al.*, 1976]. After heparin-Sepharose chromatography of solubilized acetone ether powders of dog and human liver an enzyme was eluted which closely resembled that obtained after chromatography of postheparin plasma. NaCl in concentrations of 0.5–1 M in the incubation medium activated the enzyme, but no further stimulation by preheparin plasma was observed. 24 h after hepatectomy, dog postheparin plasma was subjected to the same purification procedure. No detectable lipase activity was obtained at the expected elution volume of the NaCl gradient.

When acetone ether powders of swine and human postheparin plasma were applied to heparin-Sepharose columns and the NaCl gradient for elution of the H-TGL was extended to 1.5 M, a second activity eluted from the columns which required apolipoprotein C-II for full activity and was inhibited when assayed in the presence of 0.5–1.0 M NaCl or protamine sulfate [*Ehnholm et al.*, 1973, 1974b]. Thus, this second activity had the characteristics of LPL. These results demonstrated that affinity chromatography on heparin-Sepharose is not only a feasible procedure for purification of H-TGL but, in addition, provides so far the only satisfactory method to physically separate the two activities in postheparin plasma.

H-TGL from human postheparin plasma purified by this method had a specific activity 300 times higher than that of the original plasma [*Ehnholm et al.*, 1974b]. Maximal activity was at pH 9.0 with 0.1 and 1.0 M NaCl in the incubation medium which consisted of a sonicated triolein-gum arabic suspension in the presence of BSA. Addition of normal preheparin plasma shifted the pH optimum from 9.0 to 7.6 when the activity was assayed in 0.1 M NaCl while in 1.0 M NaCl no such change was observed. Increasing ionic strength in the assay resulted in progressively higher activity up to a maximum at 0.75 M NaCl; however, preincubation of the enzyme preparation in 1.0 M NaCl at 37 °C resulted in a considerable loss of activity which has not been noted at lower temperatures (28 °C). These observations clearly demonstrate that the activity is different from LPL, and they may explain some of the differences in pH and ionic strength optima found for lipolytic activities in rat liver perfusates and plasma membranes [*Fielding*, 1972; *Assmann et al.*, 1973a]. After gel filtration of human postheparin plasma, the activity assayed under conditions optimal for H-TGL eluted as a broad peak in the void volume. When the ionic strength of the elution buffer was increased to 0.5 M NaCl, the enzyme activity was detected as a single peak slightly before the elution volume of albumin. Purified H-TGL appeared at this volume under low- and high-salt conditions. The apparent molecular weight of the disaggregated form of the enzyme was 75,000. The data suggest aggregation of the enzyme in postheparin plasma after losing its membrane support or binding to other plasma proteins. Disruption of a large-molecular-weight complex containing the enzyme could relate to the increase in activity detected under high salt assay conditions.

Sequential use of heparin-Sepharose and concanavalin A-Sepharose affinity chromatography led to further purification of the enzyme from human postheparin plasma [*Ehnholm et al.*, 1975d]. SDS disc gel electrophoreiss revealed one major band with an estimated molecular weight of 64,000. Initially purified as a triglyceride hydrolase, the enzyme also exhibited properties of a very active phospholipase A_1 in the alkaline pH range without showing a definite decrease at pH values higher than 10. Optimal hydrolysis was obtained with equimolar concentrations of phospholipids and deoxycholate in the incubation mixture. The triglyceride hydrolase activity of the enzyme was almost identical with that of the phosphatidylethanolamine lipase activity and both were approximately 25-fold greater than the activity measured with phosphatidylcholine as substrate. This ratio of enzymatic activities was constant in postheparin plasma and during the purification procedure, and was observed even during thermal inactivation. These data strongly suggest that a single enzyme has esterolytic activity against both phospholipids and triglycerides. Therefore, it is likely that the H-TGL is principally responsible for most of the phospholipase A_1 activity found in postheparin plasma, in hepatic perfusates and in liver plasma membranes. The specificity attributed to lipolytic enzymes obviously is dependent upon the physicochemical nature of the substrate and may be determined by the environment of the lipoprotein lipids. Despite the possibility that enzymes other than H-TGL and LPL contribute to the overall activity in postheparin plasma against lipid substrates [*Nilsson-Ehle and Belfrage*, 1972], the observation that H-TGL is usually released by heparin earlier and in larger quantities than LPL and remains longer within the circulation points to a significant role of this enzyme in lipoprotein metabolism [*Brown et al.*, 1975]. Important modifications of the purification procedure for H-TGL considerably improved the yield and increased the specific activity more than 12-fold to approximately 3,000 μM free fatty acid release/h/mg enzyme protein as recently published by *Augustin et al.* [1978]. Increasing the ionic strength of the postheparin plasma before chromatography made delipidation unnecessary [*Boberg et al.*, 1977a]. Under these conditions, plasma can be directly applied to heparin-Sepharose columns with binding of H-TGL and LPL. The large recovery of the enzyme after sequential elution allows quantitative determination of both activities in human postheparin plasma as a simple and reproducible method. Introduction of isoelectric focusing as a preparative procedure resulted in further purification and, in addition, demonstrated heterogeneity within the H-TGL activity since three isoelectric points were found for enzyme preparations from several different volunteers. These subfractions may be identical with three activities which have been obtained from human postheparin plasma after formation of an enzyme substrate complex with sequential delipidation and chromatography on heparin-Sepharose [*Ehnholm et al.*, 1975b]. All three activities fulfilled H-TGL criteria; they did not require the presence of serum for full activity, were stimulated rather than inhibited by high ionic strength and shared a common antigenic site. After further purification of the fractions by calcium phosphate gel chromatography, the affinity of the enzymes for heparin is no longer retained. These data demonstrate that separation of H-TGL and LPL can only be achieved by heparin-Sepharose chromatography without prior chromatography on calcium gels. The molecular weight of H-TGL is 69,000 determined by SDS disc gel electrophoresis and gel filtration on Biogel P-100. This value is similar to size estimates of LPL from human postheparin plasma [*Fielding*, 1970; *Augustin et al.*, 1978], bovine milk [*Egelrud and Olivecrona*, 1972], swine adipose tissue [*Bensadoun et al.*, 1974], rat [*Twu et al.*, 1975] and swine heart muscle [*Ehnholm et al.*, 1975c], but differs considerably from values obtained for LPL from rat postheparin plasma [*Fielding et al.*, 1974] and rat heart [*Chung and Scanu*, 1977]. It has recently been reported that another LPL is released by heparin into rat plasma which has a molecular weight of 69,000 [*Fielding et al.*, 1977]. Amino acid composition, terminal amino acids and tryptic peptide mapping are similar, if not identical, to LPL [*Augustin et al.*, 1978]. Other investigators, however, have recently demonstrated significant differences in amino acid composition between H-TGL and LPL [*Östlund-Lindqvist*, 1978]. The purification procedure in these experiments included affinity chromatography on heparin-Sepharose with low affinity for antithrombin and SDS-polyacrylamide gel electrophoresis. H-TGL is a glycoprotein containing about 3% glucosamine, traces of galactosamine, 2% mannose and galactose and 1% sialic acid. Neither glucuronic nor iduronic acid were detected. Thus it is unlikely that the enzyme contains any heparin as a prosthetic group. In its highly purified stage, the activity is neither activated nor inhibited on addition of heparin or protamine sulfate to the incubation system. The variable effects of these substances on rather crude enzyme preparations may therefore be the result of unspecific interactions with the enzyme. Comparison of H-TGL and LPL from human postheparin plasma reveals two very similar glycoproteins, the principal difference being the carbohydrate composition, although slight deviations in the amino acid composition between the two activities cannot be ruled out. The data even suggest the possibility of interconversion of one enzyme into the other by means of glycosylation differences, alterations in the tertiary structure of the polypeptide chains, phosphorylation or peptide cleavage of a small segment. One of these mechanisms may provide selective affinity for receptor sites at the capillary endothelium of different tissues. In addition, they may explain the substantial differences which have been found between the enzymes by kinetic criteria. These include different antigenic determinants, requirement of LPL for apolipoprotein C-II as a cofactor for hydrolysis of triglycerides and different sensitivities to changes in ionic strength or the presence of protamine sulfate. Another different feature is the sensitivity to detergents [*Baginsky and Brown*, 1977]. SDS at concentrations of 1 mM in the assay system completely inactivated H-TGL but produced little or no inhibition of LPL. Conversely, Triton X 100 and sodium taurodeoxycholate protected and activated H-TGL and had no effect on LPL. Detergent-like effects have also been reported for palmitoyl-CoA [*Taketa and Pogell*, 1966]. The hydrolysis rate of this substrate is different in rat and human as compared to triglyceride [*Jansen and Hülsman*, 1974; *Baginsky and Brown*, 1977]. The results strongly suggest that comparison of H-TGL from different animals and from the human is complicated, if not possible, and can only be done under strictly identical assay conditions. Differential characteristics of the enzyme in these species may also provide the basis for different physiological roles of H-TGL.

Techniques for Selective Measurement of H-TGL and LPL from Postheparin Plasma

Four different assays have recently been developed which allow the determination of plasma H-TGL and LPL. One assay is based on selective inactivation of extrahepatic lipase by protamine sulfate [*Krauss et al.*, 1974]. LPL, derived from extrahepatic sources is determined as protamine-inactivated lipase activity; hepatic lipase activity is protamine-resistant under the conditions of the particular assay. Separation of the two enzymes by affinity chromatography on small columns with heparin covalently bound to Sepharose has also successfully been used for assessment of separate lipase activity [*Boberg et al.*, 1977a]. Probably the most specific assay method for the two enzymes is based on the use of antiserum prepared against purified plasma H-TGL. Hepatic lipase is inactivated by enzyme-antibody precipitation and subsequent assay of the remaining triglyceride lipase activity is performed at optimal conditions for LPL [*Huttunen et al.*, 1975; *Greten et al.*, 1976]. In view of the wide clinical interest for individual enzyme determination, these three methods have recently been compared [*Greten et al.*, 1977]. Good correlations were found between the immunochemical and the heparin-Sepharose method but not between these and the protamine sulfate inhibition procedure. However, in another study, significant correlations were present between the activities of protamine-sensitive lipase and immunochemically determined LPL on the one hand and between protamine-resistant lipase and H-TGL on the other [*Huttunen et al.*, 1975]. Recently, it could be shown that the very dissimilar effects of detergents on preparations of H-TGL and LPL may form the basis for the direct assay of each enzyme in the presence of the other [*Baginsky and Brown*, 1977]. SDS at a concentration of 1 mM produced no inhibition of LPL while completely inactivating H-TGL.

H-TGL in Clinical and Pharmacological Studies

It has been inferred that low PHLA in clinical disorders associated with or without hypertriglyceridemia reflects a deficiency of LPL. Low LPL in turn may then be related to a defect in lipoprotein catabolism. With methods available that distinguish between LPL and H-TGL these studies have recently been reinvestigated. Interesting results have been obtained by different investigators which suggest possible specificity of function of the enzymes in lipoprotein metabolism. The evaluation of possible differences in substrate specificity between H-TGL and LPL, however, is complicated by several factors. Pure enzyme preparations of postheparin plasma are difficult to standardize and incubation with lipoproteins *in vitro* do not necessarily reflect the activities of these enzymes *in situ*. The preferred hydrolysis of chylomicron and very-low-density lipoprotein lipids by one or the other lipase as well as the hydrolysis of these lipoproteins by both enzymes at similar rates has been shown by different research groups [*Fielding et al.*, 1977; *Krauss et al.*, 1973; *Baginsky et al.*, 1975; *Augustin et al.*, 1976]. At present, the nature of the reaction catalyzed by plasma hepatic lipase *in vivo* has not been elucidated. Yet, in several clinical studies, selective deficiency of hepatic lipase has been demonstrated in parallel with increased lipoprotein triglyceride concentration. Specific changes have also been detected with hyperlipidemic diets, drugs or hormones leading to altered plasma H-TGL activity. Studies of individual lipase activity – heparin released and *in situ* – will serve for better understanding of the respective role H-TGL plays in lipoprotein catabolism.

Clinical Disorders

Determination of H-TGL and LPL has been performed in plasma from diabetic patients and sex-matched nondiabetic control subjects [*Nikkilä et al.*,

1977a]. While LPL was significantly decreased in untreated ketotic diabetes H-TGL activity was similar in diabetics and controls. Only in patients with hypertriglyceridemic maturity onset diabetes mean H-TGL was higher than the corresponding control group. Though H-TGL activity was not related to triglyceride removal, a significant correlation to very-low-density lipoprotein triglyceride production rate could be established in this study.

Selective lipase measurements have also been carried out in patients with hyperlipoproteinemia and clinical and biochemical evidence of hypothyroidism [*Krauss et al.*, 1974]. Significant reduction of H-TGL activity was found with normal LPL when the protamine sulfate inhibition assay was used to distinguish between the enzymes.

Severe liver disease is among other lipid and lipoprotein abnormalities often accompanied by hypertriglyceridemia. This secondary metabolic disorder has been attributed to the presence of a triglyceride-rich low-density lipoprotein which may represent some intermediate particles originating from very-low-density lipoproteins and chylomicrons [*Müller et al.*, 1974]. *Klose et al.* [1977] recently reported on enzyme analyses which were performed in a study of patients with chronic active or persistent hepatitis and patients with cirrhosis of the liver. Total PHLA decreased with the severity of liver dysfunction and this decrease was due to low H-TGL and only to some degree to low LPL. With improvement of the disease, hypertriglyceridemia disappeared with concomitant increase of H-TGL and LPL. Similar results have been reported by *Sanar et al.* [1976].

Hyperlipidemia is commonly seen in patients with renal insufficiency. In patients with chronic uremia and in patients undergoing hemodialysis, type IV hyperlipoproteinemia is the usual pattern observed whereas the type II pattern is more common after renal transplantation. Studies on the course of this secondary lipid disorder have provided different explanations. Recently, *Mordasini et al.* [1977] performed selective immunochemical measurement of H-TGL and LPL in postheparin plasma in patients with conservatively treated severe renal insufficiency, under hemodialysis and after renal transplantation. A selective decrease of H-TGL with normal LPL was found in conservatively treated chronic uremia and in hemodialyzed patients. Elevated levels of very-low-density lipoproteins and increased triglycerides in low-density lipoproteins occurred concomitantly in these patients. In contrast, both enzymes were normal in allograft recipients. From this study, it may well be speculated that the accumulation of triglyceride-rich low-density lipoproteins as observed in the majority of patients with renal failure is a consequence of low H-TGL activity in these patients. The same authors found normal H-TGL but low

LPL activities in those patients with severe nephrotic syndrome who had both increased low-density and very-low-density lipoprotein plasma concentrations [*Mordasini et al.*, 1978].

Selective measurement of the two plasma lipases has also been applied to two large studies of patients with primary hyperglyceridemia. *Krauss et al.* [1974] demonstrated marked reduction of the protamine-inactivated lipase activity in 12 patients with hyperchylomicronemia including 5 type I patients whereas protamine-resistant lipase was below normal only in 1. Mean values for both enzymes were not reduced in 32 other patients also having hyperchylomicronemia, 9 with type III hyperlipoproteinemia and 23 with type IV hyperlipoproteinemia. With the enzyme-antibody precipitation technique, *Greten et al.* [1976] showed that 8 patients with type I and 2 of 4 patients with type V lipoprotein pattern had extremely low LPL levels with normal values for H-TGL. This finding implies heterogeneity among patients with type V hyperlipoproteinemia. Normal lipases were found in the patients with floating β-lipoproteinemia as they were in 6 patients with familial lecithin-cholesteryl-acyl-tansferase deficiency. The accumulation of a triglyceride-rich low-density lipoprotein in the latter group of patients, therefore, cannot be explained by decreased levels of either H-TGL or LPL in postheparin plasma.

Effect of Hormones and Drugs on Human H-TGL

Several steroid hormones have been shown to elevate triglyceride levels in plasma and may sometimes lead to severe hypertriglyceridemia with pancreatitis. It has been inferred that this rise in triglyceride concentration following estrogen administration is a consequence of impaired clearance because of suppressed PHLA. In contrast to low PHLA, several studies of triglyceride turnover with different techniques have not revealed impaired triglyceride removal during estrogen therapy. In an attempt to clarify this apparent discrepancy, *Applebaum et al.* [1977] have selectively measured H-TGL and LPL in postheparin plasma from women during estrogen therapy. This study demonstrated that the decline in PHLA is attributable to low H-TGL. The change in hepatic lipase correlated with the change in PHLA, however, it did not correlate with the increase in triglyceride concentration. Studies by *Glueck et al.* [1976] also indicated an estrogen-associated depression in protamine-resistant H-TGL.

Steroids with anabolic and progestational activity lowered plasma triglyceride concentration possibly by improving plasma very-low-density lipoprotein removal. In a recent study, *Ehnholm et al.* [1975a] were able to show that H-TGL activities selectively increased following oxandrolone treatment. However, despite the kinetic studies by *Glueck et al.* [1973], who demonstrated increased triglyceride removal after oxandrolone administration, no correlation between drug-induced lipase activity and plasma triglyceride levels could be established. Several investigators have also studied the effect of clofibrate on H-TGL and LPL in normals and hypertriglyceridemic patients [*Ganesan et al.*, 1976; *Boberg et al.*, 1977; *Greten et al.*, 1977; *Nikkilä et al.*, 1977b]. It was generally found that both in normals and patients with type IV hyperlipoproteinemia clofibrate treatment leads to a specific increase of plasma LPL while H-TGL activity remains normal. The clofibrate analogue bezafibrate, however, increased both H-TGL and LPL in normal postheparin plasma [*Greten*, in preparation, 1979]. The lack of correlation between increased or decreased enzyme activity and plasma triglyceride concentration is not fully understood at present. Drug or hormone administration may just lead to unspecific hepatic enzyme reduction and may have no relation to plasma triglyceride metabolism. Measurement of the membrane-bound enzyme after intravenous injection of heparin also may not reflect actual enzyme tissue activity.

Concluding Remarks

The importance of the liver in lipoprotein metabolism has been well documented, although its exact role in the degradation of lipoproteins is not fully understood. The liver is the major site of removal of chylomicron cholesterol [*Goodman*, 1962; *Quarfordt and Goodman*, 1967, 1969] whereas this organ accounts for only 20–30% of chylomicron triglyceride removal, the remainder is metabolized by extrahepatic tissues [*Olivecrona and Belfrage*, 1965; *Ontko and Zilversmit*, 1967; *Bergman et al.*, 1971]. In functionally hepatectomized animals, chylomicron triglyceride was cleared and the cholesteryl-ester-rich particles remained in the circulation [*Nestel et al.*, 1962; *Mayes and Felts*, 1967; *Redgrave*, 1970]. These experiments indicate that chylomicrons are initially processed in extrahepatic tissues by the action of LPL where about 70–80% of the triglycerides are removed. The cholesteryl-ester-rich remnants could then be returned to the liver for final degradation [*Noel et al.*, 1975]. However, analysis of these remnants in hepatectomized rats indicated that these particles still consist of 80% triglycerides and their size of 800 Å is similar to large very-low-density lipoproteins. Therefore, even if a large portion of chylomicron triglycerides were catabolized in extrahepatic tissues, degradation of remnants would still require a considerable lipolytic potential. In addition, the liver has been shown to remove unhydrolyzed exogenous triglyceride from the blood which is trapped in the extravascular spaces of the organ [Schotz et al., 1966]. Endogenous triglyceride is also removed from the blood and rapidly incorporated into liver esters. The peculiar structure of the hepatic microcirculation has been considered relevant for direct contact of lipoproteins with the microvillous surface of the hepatocyte via the space of Disse [Stein and Stein, 1967]. At this location, the enzyme H-TGL may hydrolyze triglycerides from remnants and very-low-density lipoproteins. If H-TGL acted as a membrane-bound enzyme it might facilitate very-low-density lipoprotein and chylomicron remnant binding to the plasma membrane with subsequent internalization [Chajek et al., 1977]. Recognition of remnants by specific receptors due to bound LPL has also been suggested [Felts et al., 1975]. The location at the plasma membrane would also explain why H-TGL is usually released by heparin into plasma earlier and in larger quantities than LPL. However, the heparin-releasable portion of the enzyme is not necessarily identical with the amount of activity having direct access to triglyceride-rich lipoproteins. Since chylomicron triglycerides are completely hydrolyzed by H-TGL *in vitro* [Augustin et al., 1976], enzyme release into the circulation during the process of intravascular lipoprotein degradation may also occur. The physiological significance of phospholipase A and CoA-ester hydrolase activity of H-TGL is unknown, especially the latter is not a common constituent of plasma lipoproteins.

The exact role of the lipase activities in the different cell compartments is not fully understood. By kinetic criteria, only lysosomal acid lipase is completely different from the lipase activities found in other various organelles. This enzyme may serve for final lysosomal lipid breakdown. Assignment of a specific role to the other activities in cell physiology is still uncertain. In addition to removal of intracellular lipids, they may be involved in transacylation reactions providing the hepatocyte with particular phospholipid species. These could then induce localized micellar reorganization of intramembranal lipids, which would favor the formation of fusion areas between subcellular membranes within the cell and thereby facilitate exchange of material between these membranes [Nachbaur et al., 1972]. In view of the finding that H-TGL and LPL may share a common polypeptide structure with the principal differences found in the carbohydrate components [Augustin et al., 1978], it is tempting to suggest the same feature for some of the lipases in the different subcellular fractions of liver. This is supported by the finding that these enzymes, except lysosomal lipase, possess similar kinetic characteristics. Different degrees of glycosylation may determine specific receptor binding and substrate affinity. There may be even a precusor-product relationship between one of these activities and the enzyme located at the plasma membrane.

The functional role of hepatic lipases in plasma or cell lipoprotein metabolism is not clear at present and can only be speculated upon. The recent data on different plasma triglyceride lipase activities, measured in clinical disorders accompanied with hypertriglyceridemia, however, favors the hypothesis of a specific function of LPL and H-TGL.

Acknowledgements

The experiments of this review were supported in part by the Deutsche Forschungsgemeinschaft, Sonderforschungsbereich 90, Cardiovasculäres System. We would like to thank Mrs. K.Jurutka for her very valuable assistance during the preparation of this manuscript.

References

Anderson, N.G.; Lansing, A.I.; Lieberman, I.; Rankin, C.T., jr., and Elrod, J.: Isolation of plasma membranes from rat liver. Wistar Inst. Monogr. *8:* 23–30 (1968).

Applebaum, D.M.; Goldberg, A.P.; Pykälistö, O.J.; Brunzell, J., and Hazzard, W.R.: Effect of estrogen on post-heparin lipolytic activity. Selective decline in hepatic triglyceride lipase. J. clin. Invest. *59:* 601–608 (1977).

Assmann, G.; Krauss, R.M.; Fredrickson, D.S., and Levy, R.I.: Characterization, subcellular localization, and partial purification of a heparin-released triglyceride lipase from rat liver. J. biol. Chem. *248:* 1992–1999 (1973a).

Assmann, G.; Krauss, R.M.; Fredrickson, D.S., and Levy, R.I.: Positional specificity of triglyceride lipases in post-heparin plasma. J. biol. Chem. *248:* 7184–7190 (1973b).

Augustin, J.; Freeze, H.; Tejada, P., and Brown, W.V.: A comparison of molecular properties of hepatic triglyceride lipase and lipoprotein lipase from human post-heparin plasma. J. biol. Chem. *253:* 2912–2920 (1978).

Augustin, J.; Geursen, R.; Klose, G., and Greten, H.: Degradation of chylomicrons with purified human plasma hepatic triglyceride lipase and lipoprotein lipase. Abstract No. 131. 10th A. Meet. Eur. Soc. Clin. Invest., 1976.

Baginsky, M.L.; Augustin, J., and Brown, W.V.: A comparative study of two human heparin-releasable enzymes each with thioesterase and lipase activities. Abstract. 1975 Pacific Slope Biochem. Conf., Honolulu 1975.

Baginsky, M.L. and Brown, W.V.: Differential characteristics of purified hepatic triglyceride lipase and lipoprotein lipase from human post-heparin plasma. J. Lipid Res. *18:* 423–437 (1977).

Beaudet, A.L.; Lipson, M.H.; Ferry, G.D., and Nicols, B.L., jr.: Acid lipase in cultured fibroblasts: cholesterol ester storage disease. J. Lab. clin. Med. *84:* 54–61 (1974).

Bensadoun, A.C.; Ehnholm, C.; Steinberg, D., and Brown, W.V.: Purification and characterization of lipoprotein lipase from pig adipose tissue. J. biol. Chem. *249:* 2220–2227 (1974).

Bensadoun, A. and Koh, T.L.: Identification of an adipose tissue-like lipoprotein lipase in perfusates of chicken liver. J. Lipid Res. *18:* 768–773 (1977).

Bergman, E.N.; Havel, R.J.; Wolfe, B.M., and Bøhmer, T.: Quantitative studies of the metabolism of chylomicron triglycerides and cholesterol by liver and extrahepatic tissues of sheep and dogs. J. clin. Invest. *50:* 1831–1839 (1971).

Biale, Y.; Gorin, E., and Shafrir, E.: Characterization of tissue lipolytic and esterolytic activities cleaving full and partial glycerides. Biochim. biophys. Acta *152:* 28–39 (1968).

Bjørnstad, P.: Phospholipase activity in rat liver microsomes studied by the use of endogenous substrates. Biochim. biophys. Acta *116:* 500–510 (1966).

Boberg, J.; Augustin, J.; Baginsky, M.L.; Tejada, P., and Brown, W.V.: Quantitative determination of hepatic and lipoprotein lipase activities from human post-heparin plasma. J. Lipid Res. *18:* 544–547 (1977a).

Boberg, J.; Boberg, M.; Gross, R.; Grundy, S.; Augustin, J., and Brown, W.V.: The effect of treatment with clofibrate on hepatic triglyceride and lipoprotein lipase activities of post-heparin plasma in male patients with hyperlipoproteinemia. Atherosclerosis *27:* 499–503 (1977b).

Boberg, J.; Carlson, L.A., and Normell, L.: Production of lipolytic activity by the isolated perfused dog liver in response to heparin. Life Sci. *3:* 1011–1019 (1964).

Brown, W.V.; Baginsky, M.L.; Boberg, J., and Augustin, J.: Lipases and lipoproteins; in Greten, Lipoprotein metabolism, pp. 2–6 (Springer, Berlin 1975).

Brown, W.V.; Levy, R.I., and Fredrickson, D.S.: Further characterization of apolipoproteins from the human plasma very low density lipoproteins. J. biol. Chem. *245:* 6588–6594 (1970).

Carter, J.R., jr.: Hepatic lipase in the rat. Biochim. biophys. Acta *137:* 147–156 (1967).

Chajek, T.; Friedman, G.; Stein, O., and Stein, Y.: Effect of colchicine, cycloheximide and chloroquine on the hepatic triacylglycerol hydrolase in the intact rat and perfused liver. Biochim. biophys. Acta *488:* 270–279 (1977).

Chung, J. and Scanu, A.M.: Isolation, molecular properties, and kinetic characterization of lipoprotein lipase from rat heart. J. biol. Chem. *252:* 4202–4209 (1977).

Claycomb, W.C. and Kilsheimer, G.S.: Purification and properties of a lipase from rat liver mitochondria. J. biol. Chem. *246:* 7139–7143 (1971).

Coleman, R.; Mitchell, R.H.; Finean, J.B., and Hawthorne, J.N.: Purified plasma membrane fraction isolated from rat liver under isotonic conditions. Biochim. biophys. Acta *135:* 573–579 (1967).

Condon, R.E.; Tobias, H., and Datta, D.V.: The liver and post-heparin plasma lipolytic activity in dog and man. J. clin. Invest. *44:* 860–869 (1965).

Datta, D.V. and Wiggins, H.S.: New effects of sodium chloride and protamine on human post-heparin plasma 'lipoprotein lipase' activity. Proc. Soc. exp. Biol. Med. *115:* 788–792 (1964).

De Duve, C.; Pressman, B.C.; Gianetto, R.; Wattiaux, R., and Appelmans, F.: Tissue fractionation studies. 6. Intracellular distribution pattern of enzymes in rat liver tissue. Biochem. J. *60:* 604–617 (1955).

De Haas, G. H.; Sarda, L., and Roger, J.: Positional specific hydrolysis of phospholipids by pancreatic lipase. Biochim. biophys. Acta *106:* 638–640 (1965).

De Jong, J. N. G.; van den Besselaar, A. M. H. P., and van den Bosch, H.: Studies on lysophospholipases. VIII. Immunochemical differences between two lysophospholipases from beef liver. Biochim. biophys. Acta *441:* 221–230 (1976).

De Jong, J. N. G.; van den Bosch, H.; Rijken, D., and Van Deenen, L. L. M.: Studies on lysophospholipases. III. The complete purification of two proteins with lysophospholipase activity from beef liver. Biochim. biophys. Acta *369:* 50–63 (1974).

Egelrud, T. and Olivecrona, T.: The purification of a lipoprotein lipase from bovine skim milk. J. biol. Chem. *247:* 6212–6217 (1972).

Egelrud, T. and Olivecrona, T.: Purified bovine milk (lipoprotein) lipase: activity against lipid substrates in the absence of exogenous serum factors. Biochim. biophys. Acta *306:* 115–127 (1973).

Ehnholm, C.; Bensadoun, A., and Brown, W. V.: Separation and partial purification of two types of triglyceride lipase from swine post-heparin plasma. J. clin. Invest. *52:* 26a (1973).

Ehnholm, C.; Greten, H., and Brown, W. V.: A comparative study of post-heparin lipolytic activity and a purified human plasma triacylglycerol lipase. Biochim. biophys. Acta *360:* 68–77 (1974a).

Ehnholm, C.; Huttunen, J. K.; Kinnunen, P. K. J.; Miettinen, T. A., and Nikkilä, E. A.: Effect of oxandrolone on post-heparin plasma lipases. New Engl. J. Med. *292:* 1314–1317 (1975a).

Ehnholm, C.; Kinnunen, P. K. J., and Huttunen, J. K.: Heterogeneity of salt-resistant lipase from human post-heparin plasma. FEBS Lett. *52:* 191–194 (1975b).

Ehnholm, C.; Kinnunen, P. K. J.; Huttunen, J. K.; Nikkilä, E. A., and Ohta, M.: Purification and characterization of lipoprotein lipase from pig myocardium. Biochem. J. *149:* 649–655 (1975c).

Ehnholm, C.; Shaw, W.; Greten, H., and Brown, W. V.: Purification from human plasma of a heparin-released lipase with activity against triglyceride and phospholipids. J. biol. Chem. *250:* 6756–6761 (1975d).

Ehnholm, C.; Shaw, W.; Greten, H.; Lengfelder, W., and Brown, W. V.: Separation and characterization of two triglyceride lipase activities from human post-heparin plasma; in Schettler and Weizel, Atherosclerosis III, pp. 557–561 (Springer, Berlin 1974b).

Felts, J. M.; Itakura, H., and Crane, R. T.: The mechanism of assimilation of constituents of chylomicrons, very low density lipoproteins and remnants – A new theory. Biochem. biophys. Res. Commun. *66:* 1467–1475 (1975).

Felts, J. M. and Mayes, P. A.: Release of lipoprotein lipase from the perfused liver of the rat. Nature, Lond. *214:* 620–621 (1967).

Fielding, C. J.: Human lipoprotein lipase. I. Purification and substrate specificity. Biochim. biophys. Acta *206:* 109–117 (1970).

Fielding, C. J.: Further characterization of lipoprotein lipase and hepatic post-heparin lipase from rat plasma. Biochim. biophys. Acta *280:* 569–578 (1972).

Fielding, P. E.; Shore, V. G., and Fielding, C. J.: Lipoprotein lipase: properties of the enzyme isolated from post-heparin plasma. Biochemistry, N.Y. *13:* 4318–4323 (1974).

Fielding, P. E.; Shore, V. G., and Fielding, C. J.: Lipoprotein lipase. Isolation and characterization of a second enzyme species from post-heparin plasma. Biochemistry, N.Y. *16:* 1896–1900 (1977).

Fowler, S. and De Duve, L.: Digestive activity of lysosomes. III. The digestion of lipids by extracts of rat liver lysosomes. J. biol. Chem. *244:* 471–481 (1969).

Franson, R.; Waite, M., and LaVia, M.: Identification of phospholipase A_1 and A_2 in the soluble fraction of rat liver lysosomes. Biochemistry, N.Y. *10:* 1942–1946 (1971).

Fredrickson, D.S.; Ono, K., and Davis, L.L.: Lipolytic activity of post-heparin plasma in hypertriglyceridemia. J. Lipid Res. *4:* 24–33 (1963).

Ganesan, G.; Ganesan, D., and Bradford, R.H.: Characterization of a triglyceride hydrolase secreted by canine liver maintained *in vitro*. Proc. Soc. exp. biol. Med. *151:* 390–394 (1976a).

Ganesan, D.; Whayne, T.F., and Vitiak, J.R.: The effect of clofibrate and analogs on human post-heparin plasma lipases. Artery *2:* 467–479 (1976b).

Glueck, C.J.; Ford, S.R., and Steiner, P.: Triglyceride removal efficiency and lipoprotein lipase. Metabolism *22:* 807–814 (1973).

Glueck, C.J.; Gartside, P.; Fallat, R.W., and Mendoza, S.: Effect of sex hormones on protamine-inactivated and -resistant post-heparin plasma lipases. Metabolism *25:* 625–632 (1976).

Goldstein, J.L.; Dana, S.E.; Faust, J.R.; Beaudet, A.L., and Brown, M.S.: Role of lysosomal acid lipase in the metabolism of plasma low density lipoprotein. J. biol. Chem. *250:* 8487–8495 (1975).

Goodman, D.S.: The metabolism of chylomicron cholesterol ester in the rat. J. clin. Invest. *41:* 1886–1896 (1962).

Graham, D.M.; Lyon, R.P., and Gofman, J.W.: Blood lipids and human atherosclerosis. II. Influence of heparin upon lipoprotein metabolism. Circulation *4:* 666–673 (1951).

Greten, H.: Post-heparin plasma phospholipase in normals and patients with hyperlipoproteinemia. Klin. Wschr. *50:* 39–41 (1972).

Greten, H.; DeGrella, R.; Klose, G.; Rascher, W.L.; De Gennes, J.L., and Gjone, E.: Measurement of two plasma triglyceride lipases by an immunochemical method: studies in patients with hypertriglyceridemia. J. Lipid Res. *17:* 203–210 (1976).

Greten, H.; Laible, V.; Zipperle, G., and Augustin, J.: Comparison of assay methods for selective measurement of plasma lipase. The effect of clofibrate on hepatic and lipoprotein lipase in normals and patients with hypertriglyceridemia. Atherosclerosis *26:* 563–572 (1977).

Greten, H.; Levy, R.I.; Fales, H., and Fredrickson, D.S.: Hydrolysis of diglyceride and glyceryl-monoester diether with lipoprotein lipase. Biochim. biophys. Acta *210:* 39–45 (1970).

Greten, H.; Levy, R.I., and Fredrickson, D.S.: A further characterization of lipoprotein lipase. Biochim. biophys. Acta *164:* 185–194 (1968).

Greten, H.; Levy, R.I., and Fredrickson, D.S.: Evidence for separate monoglyceride hydrolase and triglyceride lipase in post-heparin human plasma. J. Lipid Res. *10:* 326–330 (1969).

Greten, H.; Sniderman, A.D.; Chandler, J.G.; Steinberg, D., and Brown, W.V.: Evidence for the hepatic origin of a canine post-heparin plasma triglyceride lipase. FEBS Lett. *42:* 157–160 (1974).

Greten, H.; Walter, B., and Brown, W.V.: Purification of human post-heparin plasma triglyceride lipase. FEBS Lett. *27:* 306–310 (1972).

Guder, W.; Weiss, L., and Wieland, O.: Triglyceride breakdown in rat liver. The demonstration of three different lipases. Biochim. biophys. Acta *187:* 173–185 (1969).

Gustafson, A.; Alaupovic, P., and Furman, R.H.: Studies of the composition and structure of serum lipoproteins. Separation and characterization of phospholipid-protein residues obtained by partial delipidation of very low density lipoproteins of human serum. Biochemistry, N.Y. *5:* 632–640 (1966).

Hamilton, R.L., jr.: Post-heparin plasma lipase from the hepatic circulation; University microfilms, Ann Arbor (1964).

Harland, W.R.; Winesett, P.S., and Wasserman, A.J.: Tissue lipoprotein lipase in normal individuals and in individuals with exogenous hypertriglyceridemia and the relationship of this enzyme to assimilation of fat. J. clin. Invest. *46:* 239–247 (1967).

Havel, R.J. and Gordon, R.S.: Idiopathic hyperlipemia: metabolic studies in an affected family. J. clin. Invest. *39:* 1777–1790 (1960).

Havel, R.J.; Kane, J.; Belasse, E.O.; Segel, N., and Basso, L.V.: Splanchnic metabolism of free fatty acids and production of triglycerides of very low density lipoproteins in normotriglyceridemic and hypertriglyceridemic humans. J. clin. Invest. *49:* 2017–2035 (1970a).

Havel, R.J.; Shore, V.G.; Shore, B., and Bier, D.M.: Role of specific glycopeptides of human serum lipoprotein in the activation of lipoprotein lipase. Circulation Res. *27:* 595–600 (1970b).

Hayase, K. and Tappel, A.L.: Microsomal esterase of rat liver. J. biol. Chem. *244:* 2269–2274 (1969).

Hayase, K. and Tappel, A.L.: Specificity and other properties of lysosomal lipase of rat liver. J. biol. Chem. *245:* 169–175 (1970).

Higgins, J.A. and Green, C.: The uptake of lipids by rat liver cells. Biochem. J. *99:* 631–639 (1966).

Higgins, J.A. and Green, C.: Properties of a lipase of rat liver parenchymal cells. Biochim. biophys. Acta *144:* 211–220 (1967).

Huttunen, J.K.; Ehnholm, C.; Kinnunen, P.K., and Nikkilä, E.A.: An immunochemical method for the selective measurement of two triglyceride lipases in human postheparin plasma. Clinica chim. Acta *63:* 335–347 (1975).

Jansen, J. and Hülsman, W.C.: Long-chain acyl-CoA hydrolase activity in serum: identity with clearing factor lipase. Biochim. biophys. Acta *296:* 241–248 (1973).

Jansen, H. and Hülsman, W.C.: Liver and extrahepatic contributions to post-heparin serum lipase activity of the rat. Biochim. biophys. Acta *369:* 387–397 (1974).

Jansen, H. and Hülsman, W.C.: On hepatic and extrahepatic post-heparin serum lipase activities and the influence of experimental hypercortisolism and diabetes on those activities. Biochim. biophys. Acta *398:* 337–346 (1975).

Jansen, H.; Oerlemans, M.C., and Hülsman, W.C.: Differential release of hepatic lipolytic activities. Biochem. biophys. Res. Commun. *77:* 861–867 (1977).

Kariya, M. and Kaplan, A.: Effects of acidic phospholipids, nucleotides and heparin on the activity of lipase from rat liver lysosomes. J. Lipid Res. *14:* 243–249 (1973).

Kelley, J.L.; Ganesan, D.; Bass, H.B.; Thayer, R.H., and Alaupovic, P.: Effect of estrogen on triacylglycerol metabolism: inhibition of post-heparin plasma lipoprotein lipase by phosvitin, an estrogen-induced protein. FEBS Lett. *67:* 28–31 (1976).

Klose, G.; De Grella, R., and Greten, H.: A comparative study of human tissue and postheparin plasma triglyceride lipases. Atherosclerosis *25:* 175–182 (1976).

Klose, G.; Windelband, J.; Weizel, A., and Greten, H.: Secondary hypertriglyceridemia in patients with parenchymal liver disease. Eur. J. clin. Invest. 7: 557–562 (1977).

Korn, E. D.: Clearing factor, a heparin-activated lipoprotein lipase. I. Isolation and characterization of the enzyme from normal rat heart. J. Biochem. 215: 1–14 (1955).

Krauss, R. M.; Levy, R. I., and Fredrickson, D. C.: Selective measurement of two lipase activities in post-heparin plasma from normal subjects and patients with hyperlipoproteinemia. J. clin. Invest. 54: 1107–1124 (1974).

Krauss, R. M.; Windmüller, H. G.; Levy, R. I., and Fredrickson, D. S.: Selective measurement of two different triglyceride lipase activities in rat post-heparin plasma. J. Lipid Res. 14: 286–295 (1973).

LaRosa, J. C.; Levy, R. I.; Brown, W. V., and Fredrickson, D. S.: Changes in high density lipoprotein composition after heparin-induced lipolysis. Am. J. Physiol. 220: 785–791 (1971).

LaRosa, J. C.; Levy, R. I.; Herbert, P., and Fredrickson, D. S.: A specific apoprotein activator for lipoprotein lipase. Biochem. biophys. Res. Commun. 41: 57–62 (1970).

LaRosa, J. C.; Levy, R. I.; Windmüller, H. G., and Fredrickson, D. S.: Comparison of the triglyceride lipase of liver, adipose tissue and post-heparin plasma. J. Lipid Res. 13: 356–363 (1972).

Lindgren, F. T.; Nichols, A. V., and Freeman, N. K.: Physical and chemical composition studies on the lipoproteins of fasting and heparinized human sera. J. phys. Chem., Ithaca 59: 930–938 (1955).

Mahadevan, S. and Tappel, A. L.: Lysosomal lipases of rat liver and kidney. J. biol. Chem. 243: 2849–2854 (1968).

Mayes, P. A. and Felts, J. M.: Lack of uptake and metabolism of triglycerides of serum lipoproteins of density less than 1.006 by the perfused rat liver. Biochem. J. 105: 18c–20c (1967).

Mellors, A. and Tappel, A. L.: Hydrolysis of phospholipids by a lysosomal enzyme. J. Lipid Res. 8: 479–485 (1967).

Mellors, A.; Tappel, A. L.; Sawant, P. L., and Desai, I. D.: Mitochondrial swelling and uncoupling of oxydative phosphorylation by lysosomes. Biochim. biophys. Acta 143: 299–309 (1967).

Mordasini, R.; Frey, F.; Flury, W.; Klose, G., and Greten, H.: Selective deficiency of hepatic triglyceride lipase in uremic patients. New Engl. J. Med. 297: 1362–1366 (1977).

Mordasini, R.; Oeliker, O.; Lütschg, J., and Greten, H.: Disturbed lipid metabolism in patients with nephrosis. Abstract. 12th A. Meet. Eur. Soc. Clin. Invest., 1978.

Müller, P.; Fellin, R.; Lambrecht, J.; Agostini, B.; Wieland, H.; Rost, W., and Seidel, D.: Hypertriglyceridemia secondary to liver disease. Eur. J. clin. Invest. 4: 419–428 (1974).

Nachbaur, J.; Colbeau, A., and Vignais, P. M.: Distribution of membrane-confined phospholipases A in the rat hepatocyte. Biochim. biophys. Acta 274: 426–446 (1972).

Nachbaur, J. and Vignais, P. M.: Localization of phospholipase A_2 in outer membrane of mitochondria. Biochem. biophys. Res. Commun. 33°: 315–320 (1968).

Naito, C. and Felts, J. M.: Influence of heparin on the removal of serum lipoprotein lipase by the perfused liver of the rat. J. Lipid Res. 11: 48–93 (1970).

Nestel, P. J.; Havel, R. J., and Bezman, A.: Sites of initial removal of chylomicron triglyceride fatty acids from the blood. J. clin. Invest. 41: 1915–1921 (1962).

Neville, D. M., jr.: The isolation of a cell membrane fraction from rat liver. J. biophys. biochem. Cytol. 8: 413–422 (1960).

Newkirk, J.D. and Waite, M.: Identification of a phospholipase A_1 in plasma membranes of rat liver. Biochim. biophys. Acta 225: 224–233 (1971).

Newkirk, J.D. and Waite, M.: Phospholipid hydrolysis by phospholipases A_1 and A_2 in plasma membranes and microsomes of rat liver. Biochim. biophys. Acta 298: 562–576 (1973).

Nikkilä, E.A.; Huttunen, J.K., and Ehnholm, C.: Post-heparin plasma lipoprotein lipase in diabetes mellitus. Relationship to plasma triglyceride metabolism. Diabetes 26: 11–21 (1977a).

Nikkilä, E.A.; Huttunen, J.K., and Ehnholm, C.: Effect of clofibrate on post-heparin plasma triglyceride lipase activities in patients with hypertriglyceridemia. Metabolism 26: 179–185 (1977b).

Nilsson-Ehle, P. and Belfrage, P.: A monoglyceride hydrolyzing enzyme in human post-heparin plasma. Biochim. biophys. Acta 270: 60–64 (1972).

Nilsson-Ehle, P.; Egelrud, T.; Belfrage, P.; Olivecrona, T., and Borgström, B.: Positional specificity of purified milk lipoprotein lipase. J. biol. Chem. 248: 6734–6737 (1973).

Noel, S.P.; Dolphin, P.J., and Rubinstein, D.: An *in vitro* model for the catabolism of rat chylomicrons. Biochem. biophys. Res. Commun. 63: 764–772 (1975).

Okuda, H. and Fujii, S.: Relationship between lipase and esterase. J. Biochem., Tokyo 64: 377–385 (1968).

Olivecrona, T. and Belfrage, P.: Mechanisms of removal of chyle triglyceride from the circulating blood as studied with ^{14}C glycerol and ^3H palmitic acid labelled chyle. Biochim. biophys. Acta 98: 81–93 (1965).

Olivecrona, T.; Egelrud, T.; Iverius, P.H., and Lindahl, U.: Evidence for an ionic binding of lipoprotein lipase to heparin. Biochem. biophys. Res. Commun. 43: 524–529 (1971).

Olson, A.C. and Alaupovic, P.: Rat liver triglyceride lipase. Biochim. biophys. Acta 125: 185–187 (1966).

Ontko, J.A. and Zilversmit, D.B.: Metabolism of chylomicrons by the isolated rat liver. J. Lipid Res. 8: 90–96 (1967).

Östlund-Lindqvist, A.-M.: Lipoprotein lipase and salt resistant lipase. Purification, characterization and studies on the activation of lipases from bovine milk and human post-heparin plasma. Abstract. Acta Univ. upsal. 301: 5–28 (1978).

Patrick, A.D. and Lake, B.D.: An acid lipase deficiency in Wolman's disease. Biochem. J. 112: 1–29 (1969).

Quarfordt, S.H. and Goodman, D.S.: Metabolism of doubly labelled chylomicron cholesterol esters in the rat. J. Lipid Res. 8: 266–272 (1967).

Quarfordt, S.H. and Goodman, D.S.: Chylomicron cholesterol ester metabolism in the perfused rat liver. Biochim. biophys. Acta 176: 863–872 (1969).

Redgrave, T.G.: Formation of cholesteryl ester-rich particulate lipid during metabolism of chylomicrons. J. clin. Invest. 49: 465–471 (1970).

Sanar, J.; Blomhoff, J.R., and Gjone, E.: Triglyceride lipases in acute hepatitis. Clinica chim. Acta 71: 403–411 (1976).

Sarda, L.; Maylié, M.F.; Roger, J. et Desnuelle, P.: Comportement de la lipase pancréatique sur Séphadex. Application à la purification et à la détermination du poids moléculaire de cet enzyme. Biochim. biophys. Acta 89: 183–185 (1964).

Scherphof, G.L. and Van Deenen, L.L.M.: Phospholipase A activity of rat liver mitochondria. Biochim. biophys. Acta 98: 204–206 (1965).

Scherphof, G.L.; Waite, M., and Van Deenen, L.L.M.: Formation of lysophosphatidyl-

ethanolamines in cell fractions of rat liver. Biochim. biophys. Acta *125:* 406–409 (1966).
Schotz, M.C.; Arnesjö, B., and Olivecrona, T.: The role of the liver in the uptake of plasma and chyle triglycerides in the rat. Biochim. biophys. Acta *125:* 485–495 (1966).
Senior, J.R. and Isselbacher, K.: Demonstration of an intestinal monoglyceride lipase: an enzyme with a possible role in the intracellular completion of fat digestion. J. clin. Invest. *42:* 187–197 (1963).
Sloan, H.R. and Fredrickson, D.S.: Enzyme deficiency in cholesteryl ester storage disease. J. clin. Invest. *51:* 1923–1926 (1972).
Spitzer, J.A. and Spitzer, J.J.: Effect of liver on lipolysis by normal and post-heparin sera in the rat. Am. J. Physiol. *185:* 18–22 (1956).
Stein, O. and Stein, Y.: The role of the liver in the metabolism of chylomicrons, studied by electron microscopic autoradiography. Lab. Invest. *17:* 436–446 (1967).
Stoffel, W. and Greten, H.: Studies in lipolytic activities of rat liver lysosomes. Hoppe-Seyler's Z. physiol. Chem. *338:* 1145–1150 (1967).
Stoffel, W. and Trabert, V.: Studies on the occurrence and properties of lysosomal phospholipases A_1 and A_2 and the degradation of phosphatidic acid in rat liver lysosomes. Hoppe-Seyler's Z. physiol. Chem. *350:* 836–844 (1969).
Taketa, K. and Pogell, B.M.: The effect of palmitoyl-coenzyme A on glucose-6-phosphate dehydrogenase and other enzymes. J. biol. Chem. *241:* 720–726 (1966).
Teng, M. and Kaplan, A.: Purification and properties of rat liver lysosomal lipase. J. biol. Chem. *249:* 1064–1070 (1974).
Torquebiau-Colard, O.; Paysant, M.; Wald, R. et Polonovski, J.: Phospholipase A de membranes plasmiques isolées de foie de rat. Action sur les phospholipides exogènes. Bull. Soc. Chim. biol. *52:* 1061–1071 (1970).
Twu, J.S.; Garfinkel, A.S., and Schotz, M.: Rat heart lipoprotein lipase. Atherosclerosis *22:* 463–472 (1975).
Van den Bosch, H. and De Jong, J.G.N.: Studies on lysophospholipases. IV. The subcellular distribution of two lysolecithine hydrolyzing enzymes in beef liver. Biochim. biophys. Acta *398:* 244–257 (1975).
Victoria, E.J.; van Golde, L.M.G.; Hostetler, K.Y.; Scherphof, G.L., and Van Deenen, L.L.M.: Some studies on the metabolism of phospholipids in plasma membranes from rat liver. Biochim. biophys. Acta *239:* 443–457 (1971).
Vogel, W.C. and Bierman, E.L.: Post-heparin human lecithinase in man and its positional specificity. J. Lipid Res. *8:* 46–53 (1967).
Vogel, W.C. and Bierman, E.L.: Evidence for '*in vivo*' activity of post-heparin plasma lecithinase in man. Proc. Soc. exp. Biol. Med. *127:* 77–80 (1968).
Vogel, W.C. and Bierman, E.L.: Correlation between post-heparin lipase and phospholipase activities in human plasma. Lipids *4:* 385–395 (1969).
Vogel, W.C.; Ryan, W.G.; Kapel, J.L., and Olwin, J.H.: Post-heparin phospholipase and fatty transesterification in human plasma. J. Lipid Res. *6:* 335–340 (1965).
Vogel, W.W. and Zieve, L.: Post-heparin phospholipase. J. Lipid Res. *5:* 177–183 (1964).
Waite, M.; Scherphof, G.L.; Boshouwers, F.M.G., and Van Deenen, L.L.M.: Differentiation of phospholipases A in mitochondria and lysosomes of rat liver. J. Lipid Res. *7:* 411–420 (1966).
Waite, M. and Sisson, P.: Partial purification and characterization of the phospholipase A_2 from rat liver mitochondria. Biochemistry, N.Y. *10:* 2377–2383 (1971).

Waite, M. and Sisson, P.: Solubilization by heparin of the phospholipase A_1 from the plasma membranes of rat liver. J. biol. Chem. *248:* 7201–7206 (1973a).

Waite, M. and Sisson, P.: Utilization of neutral glycerides and phosphatidylethanolamine by the phospholipase A_1 of the plasma membranes of rat liver. J. biol. Chem. *248:* 7985–7992 (1973b).

Waite, M. and Sisson, P.: Studies on the substrate specificity of the phospholipase A_1 of the plasma membrane of rat liver. J. biol. Chem. *249:* 6401–6405 (1974).

Waite, M. and Van Deenen, L.L.M.: Hydrolysis of phospholipids and glycerides by rat-liver preparations. Biochim. biophys. Acta *137:* 498–517 (1967).

Wallach, D.H.F. and Ullrey, D.: Studies on the surface of cytoplasmic membranes of Ehrlich ascites carcinoma cells. II. Alkalization-activated adenosine triphosphate hydrolysis in a microsomal membrane fraction. Biochim. biophys. Acta *88:* 620–629 (1964).

Whayne, T.F., jr.; Felts, J.M., and Harris, P.A.: Effect of heparin on the inactivation of serum lipoprotein lipase by the liver in unanesthetized dogs. J. clin. Invest. *48:* 1246–1251 (1969).

Zieve, F.J. and Zieve, L.: Post-heparin phospholipase and post-heparin lipase have different tissue origins. Biochem. biophys. Res. Commun. *47:* 1480–1485 (1972).

Dr. J. Augustin, Klinisches Institut für Herzinfarktforschung
an der Medizinischen Universitätsklinik Heidelberg, D-6900 Heidelberg (FRG)

Lecithin: Cholesterol Acyltransferase

An Exercise in Comparative Biology[1]

John A. Glomset

Howard Hughes Medical Institute; Departments of Medicine and Biochemistry, and Regional Primate Research Center, University of Washington, Seattle, Wash.

Lecithin: cholesterol acyltransferase (LCAT) is an enzyme that is synthesized in the liver, circulates in the plasma, and acts on high-density lipoproteins (HDL) [30, 33, 35, 39, 76]. By catalyzing fatty acid transfer from the C:2 position of phosphatidylcholine to unesterified cholesterol, it promotes the formation of lysophosphatidylcholine and of cholesteryl esters that are rich in essential fatty acids. Why these lipids should be formed in the plasma is not yet fully understood. However, evidence is accumulating that the reaction may play very different roles in different species. Two viable possibilities in humans are that the reaction removes 'excess' surface phosphatidylcholine and unesterified cholesterol from remnants of chylomicrons and very-low-density lipoproteins (VLDL) and that the reaction promotes cholesterol transport from peripheral tissues to the liver. In addition, studies of patients with familial LCAT deficiency suggest that the reaction probably also affects the metabolism of apolipoproteins and other lipids. In rats, however, the reaction may be less involved in the metabolism of chylomicron and VLDL cholesterol. Instead, the available evidence suggests that it plays a role in the transport of cholesterol and essential fatty acids to the adrenals, ovaries and testes. A major objective of this chapter is to explore the basis for this species difference by comparing what is known about the LCAT reaction in humans and rats with what is known about related aspects of lipid and lipoprotein metabolism. It is hoped that emphasis on the comparative biology of the LCAT reaction will add to the understanding of lipoprotein metabolism and stimulate future research.

[1] Supported by the Howard Hughes Medical Institute and by grants HL10642, RR00166 and AG00299 from the National Institutes of Health, US Public Health Service.

Effects of the LCAT Reaction on Plasma Lipoproteins and Tissues

Three considerations should be noted throughout the following discussion. (1) Most studies of the LCAT reaction have been done with human material. (2) The most striking effects of the LCAT reaction have been observed in studies of familial LCAT deficiency. (3) Most studies of other aspects of lipid metabolism have been performed in rats. The plan is therefore to focus initially on the effects of the reaction on human lipoproteins and tissues and to interject information concerning rats where that information is available and appropriate. Later, however, the focus will necessarily be reversed.

Plasma Lipids

The effects of the LCAT reaction on plasma lipids were demonstrated by the original studies of the reaction mechanism. These studies, reviewed previously in detail [33], showed that cholesterol becomes esterified when plasma is incubated *in vitro,* and that the esterification reaction is accompanied by a nearly equimolar decrease in concentration of phosphatidylcholine. They provided evidence also that fatty acids are transferred directly from the C:2 position of phosphatidylcholine to unesterified cholesterol, forming lysophosphatidylcholine and cholesteryl esters that largely contain unsaturated fatty acids. This indicated that cholesterol esterification in the plasma is catalyzed by an acyltransferase that shows specificity for the C:2 position of phosphatidylcholine. However, the reaction is more complicated than this since about 10–12% of the fatty acids transferred are saturated, a proportion severalfold higher than would be anticipated from examination of the fatty acids in the C:2 position of phosphatidylcholine. In addition, considerably less than an equimolar amount of lysophosphatidylcholine accumulates. Finally, in human plasma the amount of arachidonic acid transferred is about half of what would be expected [34]. Since the action of LCAT on phosphatidylcholine produces 1-acyl-lysophosphatidylcholines that mainly contain saturated fatty acids, and since LCAT seems also to convert two molecules of 1-acyl-lysophosphatidylcholine to one molecule of phosphatidylcholine and one molecule of glycerylphosphorylcholine [85], it is possible that the coupled activities of the enzyme continuously generate disaturated phosphatidylcholines that act as substrates in the cholesterol esterification reaction. Although this may explain why the transfer of saturated fatty acids is unexpectedly high, it does not account for the disproportionately low transfer of arachidonic acid. This question deserves to be explored because there is a major species difference between rats and humans. The amount of arachidonic acid present in the C:2 position of plasma phosphatidylcholine in rats (as much as 65%) is much higher than in humans (as much as 25%), and arachidonic acid comprises as much as 65% of the fatty acids transferred to cholesterol in rat plasma [31, 34]. Thus, the amount of cholesteryl arachidonate formed in rats is more than fivefold greater than that formed in humans. This could reflect a major species difference in cholesteryl ester utilization which might be of importance to the metabolism of prostaglandins and longer-chain polyunsaturated fatty acids (see later).

Effects on HDL

Early studies of the LCAT reaction revealed that some human plasma lipoproteins react directly with the enzyme whereas others do not. The evidence that LCAT preferentially reacts with HDL can be summarized as follows:

(A) When human plasma cholesteryl esters are labeled by incubating plasma with radioactive cholesterol [33], or by injecting radioactive mevalonate *in vivo* [44], the individual cholesteryl esters show similar specific activities. However, the order of cholesteryl ester specific activity among the isolated lipoprotein fractions is HDL > VLDL >> low-density lipoproteins (LDL). In addition, the smaller HDL have higher specific activities than the larger HDL [36], whereas the larger VLDL have higher specific activities than the small VLDL [9]. Again, a major species difference exists between humans and rats. In rats injected with radioactive mevalonate [46], the specific activities of cholesteryl palmitate and cholesteryl oleate in the plasma are considerably higher than the specific activities of cholesteryl linoleate and cholesteryl arachidonate. Moreover, the order of cholesteryl ester specific activities among the lipoprotein fractions differs depending on the type of cholesteryl ester [29]. The specific activities of polyunsaturated esters are highest in HDL whereas those of saturated and monounsaturated cholesteryl esters are highest in VLDL. It is important to note that these differences in lipoprotein cholesteryl ester specific activity between the two species accompany differences in lipoprotein cholesteryl ester composition. The relative proportions of the various cholesteryl esters are very nearly the same in all human plasma lipoproteins [47], whereas rat VLDL cholesteryl esters are rich in cholesteryl palmitate and cholesteryl oleate compared with rat HDL, which are rich in cholesteryl arachidonate and cholesteryl linoleate [28].

(B) When human lipoprotein fractions or subfractions are first isolated from fresh plasma and then incubated with LCAT [32], HDL react actively with the enzyme, whereas LDL react only slightly, and VLDL do not react at all. In addition, HDL_3, a subfraction of HDL that contains smaller lipoproteins, reacts with LCAT, whereas HDL_2, a subfraction that contains larger lipoproteins, does not [24].

(C) Incubation studies with mixtures of purified preparations of LCAT, phosphatidylcholine and unesterified cholesterol have shown that the reaction is greatly stimulated by the presence of apolipoprotein A-I, the principal apolipoprotein component of HDL [26]. In addition, optimal activity occurs when the ratio of phosphatidylcholine to unesterified cholesterol is at least 3:1 [25,64,71]. Such high ratios occur normally in HDL but in none of the other plasma lipoproteins.

(D) Other experiments have shown that LCAT binds preferentially to HDL [3]. Thus, although the enzyme can be obtained in a highly purified form that contains little or no lipid [7], it binds to columns of HDL-Sepharose in isotonic saline at pH 7.4, and can be displaced from the columns by elution with normal HDL [5], bile acids [3] or distilled water [5], but not by elution in the presence of LDL or VLDL [5].

(E) Studies [40] of the lipoproteins of patients with familial LCAT deficiency have shown that the patients' HDL also react preferentially with LCAT, whereas the VLDL and LDL show the same lack of reactivity seen with the corresponding lipoproteins of normal plasma. These studies counter an objection that might otherwise be raised concerning the lack of reactivity of normal VLDL and LDL with LCAT. Since normal human lipoproteins all apparently contain cholesteryl esters that have been formed by the LCAT reaction, whether they react with LCAT *in vitro* or not, it could be argued that those lipoproteins that do not react with LCAT *in vitro* may have already reacted exhaustively *in vivo*. However, this objection does not apply in the case of the LDL or VLDL of patients afflicted with familial LCAT deficiency since the content of cholesteryl ester is very low in these lipoproteins, and there is no evidence of previous action of LCAT *in vivo*. Only in the case of normal HDL_2 does the argument seem to apply, since all subfractions of the patients' HDL so far obtained react with LCAT and since lipoproteins that resemble HDL_2 are formed in the course of the reaction [37].

Studies of familial LCAT deficiency have also provided evidence that the LCAT reaction strikingly affects the physical properties of HDL. HDL isolated from the patients' native plasma bear little resemblance to normal HDL except that they contain phosphatidylcholine, unesterified cholesterol, and all of the apolipoproteins that are usually found in HDL. The structures of these HDL are highly abnormal, and none of the HDL resembles normal

HDL$_2$ or HDL$_3$. Instead, there are small globular lipoproteins that contain apolipoprotein A-I, and disc-shaped lipoproteins of various sizes that contain apolipoprotein A-I and apolipoprotein A-II, with or without apoliprotein E [39,91,92]. When the patients' plasma is incubated with LCAT, these lipoproteins become converted to particles the size of normal HDL$_2$ and HDL$_3$, and the content of HDL apolipoproteins also changes. More apolipoprotein A-I and C apolipoproteins, but much less apolipoprotein E, are recovered upon preparative ultracentrifugation [37]. These changes are accompanied by events involving other lipoproteins as well (see below).

Indirect Effects of the LCAT Reaction on Other Lipoproteins

Studies of normal human plasma and of the plasma of patients with familial LCAT deficiency have yielded evidence that the action of LCAT on HDL indirectly affects both the lipid and apolipoprotein compositions of other plasma lipoproteins. As indicated earlier, when normal plasma is incubated at 37 °C, essentially equimolar changes occur in phosphatidylcholine, unesterified cholesterol and cholesteryl ester. However, equimolar changes in these lipids are not seen in separate lipoproteins isolated after the incubation [52]. Instead, the molar change in phosphatidylcholine relative to unesterified cholesterol is disproportionately high in the HDL and disproportionately low in the LDL and VLDL. In view of the evidence that LCAT acts preferentially on HDL and in view of the fact that unesterified cholesterol exchanges readily among plasma lipoproteins [75], these results seem best explained by a coupled reaction sequence that involves an initial enzymic attack on HDL followed by the nonenzymic redistribution of unesterified cholesterol from other lipoproteins to the HDL.

Studies of the lipoproteins of patients with LCAT deficiency support this formulation and strongly suggest that cholesteryl esters and apolipoproteins also redistribute. To appreciate the evidence for this, it is important to consider the abnormalities that affect the patients' native VLDL and LDL [38,39]. The VLDL are often present in abnormally high concentrations, migrate with an abnormally slow electrophoretic mobility, and contain only about half the normal amount of cholesteryl ester. The cause of the slow electrophoretic mobility seems to be an abnormally high content of apolipoprotein C-I and apolipoprotein E relative to apolipoprotein C-II, apolipoprotein C-III and apolipoprotein B [37]. The cholesteryl esters are rich in palmitate and oleate and seem to be derived from the intestinal mucosa [39].

The patients' LDL show even more striking abnormalities [39]. There are particles of the same size and appearance as normal LDL, but they contain abnormally large amounts of triglyceride relative to cholesteryl ester. Other, larger LDL resemble the 'LpX' of patients afflicted with obstructive jaundice, whereas still larger particles look like whorled bilayers under the electron microscope and mainly contain unesterified cholesterol, phosphatidylcholine and albumin [37].

When the patients' plasma is incubated with LCAT [68], both the VLDL and LDL are markedly altered. There is a major decrease in content of phosphatidylcholine and unesterified cholesterol in the abnormally large LDL, and an increase in cholesteryl ester in the VLDL and normal-sized LDL. In addition, apolipoprotein E transfers from the HDL to the VLDL, and C apolipoproteins transfer from the VLDL to the HDL. The most likely explanation of these changes is that the large and intermediate-sized LDL and the VLDL contribute phosphatidylcholine and unesterified cholesterol to HDL, where action of LCAT converts the two substrates to cholesteryl ester, and that the cholesteryl ester formed largely transfers from HDL to the normal-sized LDL and VLDL in association with apolipoprotein E. Thus, lipoproteins in the HDL fraction seem to provide a reactive surface upon which LCAT can act to form cholesteryl esters, but the lipid substrates do not necessarily originate as components of HDL, and the cholesteryl esters formed largely transfer to other lipoproteins. In addition, those cholesteryl esters that remain in the HDL seem to increase the affinity of these lipoproteins for apolipoprotein A-I, and to promote exchange of apolipoprotein E from HDL with C apolipoproteins from VLDL [37].

Although the mechanisms underlying these transfer reactions have not yet been conclusively established, it has been known for many years [19, 75] that both phosphatidylcholine and unesterified cholesterol exchange readily among plasma lipoproteins and between plasma lipoproteins and erythrocyte membranes. It has been apparent also that most of the lysophosphatidylcholine formed by the LCAT reaction either is converted to glycerylphosphorylcholine or transfers to albumin [33]. However, the fact that cholesteryl esters transfer readily among human plasma lipoproteins has been recognized only recently [2, 65, 73, 94]. Early evidence, based on studies in rats, had suggested little or no exchange of plasma lipoprotein cholesteryl esters [75]. However, *Rehnborg and Nichols* [73] subsequently demonstrated that human HDL cholesteryl esters exchange with VLDL triglycerides, and that this is promoted by the LCAT reaction. More recently still, *Zilversmit et al.* [95] and *Barter and Lally* [10] have found evidence for the existence of a cholesteryl ester exchange protein in the protein fraction of d >1.21 g/ml prepared from rabbit or human plasma. In view of the differences that have already been noted between rats and humans, it is of particular interest that rat plasma shows little or no cholesteryl ester exchange activity [10]. Whether the cholesteryl ester exchange protein is reponsible for the transfer of cholesteryl esters observed in the incubation experiments with the plasma of patients with familial LCAT deficiency has not been demonstrated. However, when the patients' plasma is incubated with LCAT, the amount of HDL cholesteryl ester that transfers to the VLDL is much larger than the amount of VLDL triglyceride that transfers to HDL [68]. Therefore, more than simple exchange of HDL core lipid for VLDL core lipid must be involved. In addition, since apolipoprotein E also transfers from the patients' HDL to their VLDL, the possibility that this apolipoprotein may act as a cholesteryl ester carrier clearly needs to be evaluated.

Effects on Cell Surfaces

Studies of human erythrocytes have provided evidence that the LCAT reaction can indirectly affect the content of cell surface lipids. *Murphy* [62] was the first to show that loss of cholesterol from erythrocyte membranes occurs when erythrocytes are incubated with plasma in the presence of LCAT. Although he did not specifically prove that the cholesterol released from the erythrocytes had become associated with plasma lipoproteins, it seemed likely that esterification of HDL cholesterol had led to a redistribution of unesterified cholesterol between the erythrocyte membranes and these lipoproteins. Support for this interpretation was subsequently obtained from studies of familial LCAT deficiency. The composition of the patients' erythrocytes was shown to be abnormal in two important respects: (1) the content of unesterified cholesterol per erythrocyte was twice that of normal erythrocytes, and (2) although the total content of phospholipid per erythrocyte was normal, the content of phosphatidylcholine was doubled at the expense of sphingomyelin and phosphatidylethanolamine [30]. These abnormalities seemed to reflect the abnormally high contents of unesterified cholesterol and phosphatidylcholine in the patients' plasma, since the content of unesterified cholesterol in the erythrocytes could be lowered by incubating the erythrocytes in normal plasma [67]. Furthermore, the content of unesterified cholesterol in normal erythrocytes could be increased by incubating the erythrocytes with the patients' plasma [67]. Finally, loss of cholesterol from the patients' HDL to normal erythrocyte membranes could be reversed by incubating the erythrocyte membranes and HDL in the presence of LCAT [40].

Although these results clearly indicate that the action of LCAT on human plasma lipoproteins influences the content of lipid in human erythrocyte membranes, it is not yet clear whether these results are generally applicable to the plasma membranes of all cells of all species. Attempts to demonstrate an effect of the LCAT reaction on various types of other cells have so far been unsuccessful [11, 66, 82], but the results of these experiments are inconclusive because the cells were not demonstrated to be in a steady state before addition of LCAT, and because equilibria between the cholesterol of plasma membranes and intracellular membranes were not considered.

Relation of the LCAT Reaction to Other Metabolic Processes

Understanding the metabolic role of the LCAT reaction in a given species clearly depends on knowledge of not only the effects of the reaction on plasma lipoproteins and tissues, but also the relation of these effects to other metabolic events. For example, it is important to relate the LCAT reaction to other reactions that produce plasma lipoprotein cholesteryl esters, and to relate the rate of the LCAT reaction to the turnover of HDL and other lipoproteins. Finally, it is important to search for relations between the LCAT reaction and other aspects of lipid transport. Although knowledge of these relations in humans and rats is still fragmentary, enough information is available to indicate that significant species differences exist.

Importance of the LCAT Reaction As a Source of Plasma Cholesteryl Esters

Evidence suggests that three potential sources of plasma cholesteryl esters vary in importance in different species: the acyl-CoA:cholesterol acyltransferase (ACAT) reaction in liver [45], the ACAT reaction in the intestinal mucosa [51] and the LCAT reaction. There are several reasons for believing that only two of these are important in humans. (1) Attempts to measure ACAT activity in human liver have been unsuccessful [83]. (2) The fatty acid compositions of the cholesteryl esters of human plasma VLDL, LDL and HDL are the same and reflect the specificity of the LCAT reaction [33]. (3) The labeling pattern of the different cholesteryl esters of human plasma VLDL, LDL and HDL following injection of radioactive mevalonate is consistent with that obtained following incubation of plasma with radioactive cholesterol *in vitro* [33]. (4) The rate of the LCAT reaction *in vitro* closely corresponds to the rate of plasma cholesteryl ester formation in fasting human subjects *in vivo* [33]. (5) The few cholesteryl esters present in the plasma of patients with familial LCAT deficiency become labeled following ingestion of radioactive cholesterol, which suggests an intestinal origin, but not following injection of radioactive mevalonate, which suggests that an ACAT is not operative in the liver [30]. It can be concluded, therefore, that human plasma cholesteryl esters are formed by ACAT in the intestine and by LCAT in the plasma, but apparently not by ACAT in the liver. In rats, however, there is incontrovertible evidence that all three reactions are operative. (1) ACAT activity can be readily demonstrated in both liver and intestine [45, 51]. (2) VLDL isolated from the Golgi of the liver clearly contain cholesteryl ester [49]. (3) The composition of rat plasma VLDL cholesteryl esters is compatible with a hepatic origin, whereas that of rat plasma HDL is compatible with formation by the LCAT reaction [28]. (4) The labeling patterns of rat plasma VLDL and HDL are different, as mentioned earlier. It has long seemed likely that this important difference between humans and rats reflects other substantial differences in lipoprotein metabolism. Additional evidence for this will be presented below.

Relation of the LCAT Reaction to the Metabolism of HDL and Other Lipoproteins

The initial rate of the LCAT reaction can be determined *in vitro* by incubating freshly obtained plasma and measuring the rate of disappearance of unesterified cholesterol [59]. Alternatively, plasma unesterified cholesterol can be labeled *in vitro* [84] or *in vivo* [55], and the rate of formation of labeled cholesteryl esters *in vitro* can be determined. Both measurements yield rates in the range of 50–100 nmol/ml/h, although the rates obtained from measurements of mass are somewhat higher than those obtained from measurements of radioactivity. Moreover, the rates calculated for humans and rats [83, 86] are quite similar.

If it is assumed that all cholesteryl esters formed by the LCAT reaction originate in HDL, then the turnover time of HDL cholesteryl esters can be approximated in humans as well as in rats[2]. These turnover times are in the same general range: 10–17 h in humans compared with about 17 h in rats. In contrast, the turnover times for HDL apolipoproteins differ appreciably in

---

[2] See 'Appendix'.

the two species. Whereas the turnover time of apolipoprotein A-I is approximately 8 days in humans [14], it is only about 12–17 h in rats [74]. This suggests that much of the cholesteryl ester formed by the action of LCAT on human HDL is catabolized by a pathway that differs from that involved in the removal of HDL apolipoprotein. In contrast, the similar turnover times of HDL cholesteryl esters and apolipoprotein A-I in rats suggest that catabolism of the components of HDL may largely occur by a common pathway in this species. A possible pathway of cholesteryl ester catabolism in humans is suggested by the incubation experiments with plasma lipoproteins of patients with familial LCAT deficiency described earlier. In these experiments 80–90% of the cholesteryl esters formed by the action of LCAT on human HDL appeared to transfer to lipoproteins that contain apolipoprotein B. It is, therefore, important to ask whether the *rates* of transfer of HDL cholesteryl esters to VLDL, intermediate-density lipoproteins (IDL), and LDL are compatible with this pathway. Although reliable estimates of the *net* rates of transfer of cholesteryl ester to these lipoproteins are not yet available, the rates of exchange of labeled cholesteryl esters among these lipoproteins have been measured [10] and found to be approximately as high as the rate of cholesteryl ester formation by LCAT. If net transfer and exchange occur by similar mechanisms, this would suggest that the rate of net transfer is not a limiting factor in human plasma. Another question that must be asked is whether the turnover times of VLDL, IDL and LDL in humans are compatible with extensive net transfer of cholesteryl esters from HDL. In humans, most VLDL are believed to be successively converted to IDL and LDL [20], and the respective turnover times of apolipoprotein B in these fractions are approximately 4–6 h [80], 6–8 h [20], and 4–5 days [56]. The cumulative turnover time of apolipoprotein-B-containing material is accordingly more than sufficient for the proposed transfer of cholesteryl ester. Indeed, the total turnover time of cholesteryl ester in apolipoprotein-B-containing lipoproteins can be calculated to be about 1.3–1.9 days, which suggests that cholesteryl esters cycle through these lipoproteins as the latter circulate in human plasma. In contrast, the situation again seems to be quite different in rats. In this species, most VLDL remnants appear to be rapidly removed from the circulation by the liver, and only a small fraction of the VLDL appears to be converted to IDL and LDL [20]. Moreover, the turnover time of VLDL in the circulation appears to be approximately 6–8 min [23] compared with 4–6 h in humans. Therefore, even if mechanisms were to exist for transferring cholesteryl esters from rat HDL to rat lipoproteins that contain apolipoprotein B, net transfer of cholesteryl esters could be of only limited significance.

Hypotheses Concerning the Role of the LCAT Reaction in Lipid Transport

Role of the Reaction in the Metabolism of Chylomicrons and VLDL

Schumacher and Adams [79] suggested several years ago that LCAT is part of a mechanism for removing excess surface lipid from remnants of chylomicrons and VLDL. Starting from the premise that the triglycerides of chylomicrons and VLDL form an interior core that is covered by a surface layer of phosphatidylcholine and unesterified cholesterol, they pointed out that a relative excess of these surface lipids should develop as triglycerides are removed by the action of lipoprotein lipase (LPL). They suggested that LCAT disposes of this excess lipid by catalyzing the conversion of phosphatidylcholine and unesterified cholesterol to cholesteryl ester, and that this facilitates further action of LPL.

Although this hypothesis needs to be modified on the basis of more recent research, there are several indications that it may be essentially valid in humans. The modifications that need to be made are based on the following considerations. (1) There is evidence that phospholipids of triglyceride-rich lipoproteins are partially hydrolysed by LPL [52]. Therefore, removal of excess phospholipid from the surfaces of these lipoproteins may be less important than removal of excess unesterified cholesterol. (2) There is no evidence yet that LCAT acts directly on either chylomicrons or VLDL or on remnants of these lipoproteins, rich in unesterified cholesterol and phosphatidyl choline. Action of LCAT on HDL can, however, indirectly lead to the removal of surface lipids from VLDL and other lipoproteins [40], so absence of direct action of LCAT on chylomicrons or VLDL is not critical. (3) There is a discrepancy between the rapid rate of removal of triglyceride from chylomicrons and VLDL [42] and the slower rate of the LCAT reaction, which suggests that LCAT does not simply act in tandem with LPL. However, this also is not critical since it appears that LCAT limits the size of a large pool of phosphatidylcholine and unesterified cholesterol in plasma lipoproteins and plasma membranes, and this may ensure the availability of binding sites for surface lipid that buffer the changes accompanying the LPL reaction. As discussed in more detail elsewhere [42], this buffering action would be effected through redistribution of the surface lipid from remnants of chylomicrons or VLDL to potential binding sites on lipoproteins and plasma membranes made available by the continuous action of LCAT on HDL. (4) Even if this redistribution is rapid relative to the rate of the LCAT reaction, the hypothesis would be valid only if redistribution were also rapid compared with the turnover time of triglyceride-rich lipoproteins in the plasma. Whereas this is probably true in humans, it is probably *not* true in rats (fig. 1). Indeed, several investigators [72] have obtained evidence that both the unesterified cholesterol and cholesteryl esters of rat chylomicron remnants are rapidly removed from the circulation by the liver. (5) In familial LCAT deficiency, the metabolism of chylomicron or VLDL triglyceride is not always impaired, which suggests that removal of excess surface phosphatidylcholine and unesterified cholesterol by the LCAT reaction does not directly affect the LPL reaction.

Despite these reservations concerning the role of the reaction, other evidence does suggest that LCAT is importantly involved in the metabolism of chylomicrons and VLDL. This evidence can be summarized as follows. (1) One of the most striking abnormalities in familial LCAT deficiency is clearly related to triglyceride metabolism. It has been suggested [41] that the large lamellar aggregates of phosphatidylcholine and unesterified cholesterol, observed among the patients' LDL, are formed from the excess surface lipids of chylomicrons and VLDL since the concentration of these aggregates decreases dramatically when the patients ingest triglyceride-free diets. Furthermore, as mentioned earlier, the concentration of these particles also decreases when the patients' plasma is incubated with LCAT *in vitro*. Both observations clearly support the hypothesis of *Schumacher and Adams* [79]. (2) Lamellar structures that resemble those observed in the plasma of patients afflicted with familial LCAT deficiency are formed when chylomicrons are incubated with LPL *in vitro* [13]. (3) The rate of the LCAT reaction in normal human plasma increases when triglyceride transport would be expected to increase [4, 59, 60, 77, 78]. An additional observation, made in chickens, also suggests that the actions of LPL and LCAT may be related since the two enzymes appear concomitantly during development [69].

Therefore, a modified hypothesis can be developed as follows. The action of LPL on chylomicrons or VLDL forms remnants that are rich in surface lipid, particularly unesterified cholesterol. If the remnants are rapidly removed from the circulation, as in rats, little action of LCAT on this lipid is required. However, in species that do not rapidly remove VLDL remnants from the circulation, but instead convert them to LDL, there is time for the excess lipid

---

Fig. 1. Schematic comparison of VLDL metabolism in humans and rats. Note that human (a) VLDL or their lipoprotein products circulate in the plasma for many hours, whereas rat (b) VLDL circulate for only a few minutes. It is proposed that whatever VLDL surface lipid accumulates in humans following the action of LPL has time to distribute

Lecithin: Cholesterol Acyltransferase

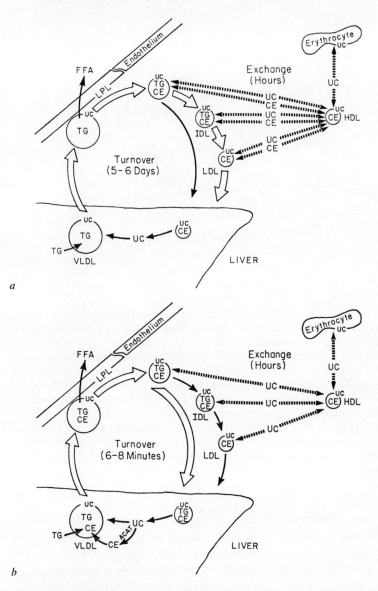

among other lipoproteins and tissues. Rat VLDL, however, is removed so quickly from the circulation that only minimal distribution occurs. The action of LCAT to remove excess surface lipid is therefore required in humans, but not in rats. In contrast, hepatic ACAT activity is required to control the content of cholesterol in rat liver but not in human liver, because of the rapid flux of rat VLDL remnants. FFA = Free fatty acids; UC = unesterified cholesterol; TG = triglycerides; CE = cholesteryl ester.

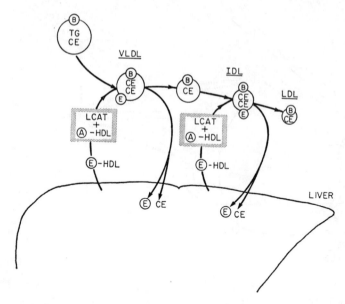

Fig. 2. Proposed apolipoprotein-E-dependent pathway for cholesteryl ester transport. This pathway for transport of cholesteryl esters in humans is based on the following observations: (1) the plasma of patients with familial LCAT deficiency contains HDL that are unusually rich in apolipoprotein E; (2) this apolipoprotein E and cholesteryl esters transfer from the HDL to VLDL when the patients' plasma is incubated with LCAT, and (3) the turnover time of plasma cholesteryl esters in normal humans is considerably shorter than that of lipoproteins containing apolipoprotein B or of HDL. It is assumed that HDL rich in apolipoprotein E are secreted by the liver, as in rats [50], that this apolipoprotein E acts as a carrier of cholesteryl esters from HDL to VLDL or from VLDL to the liver, and that this process occurs approximately two times during the life of an apolipoprotein-B-containing lipoprotein, i.e., during the conversion of VLDL to LDL. TG = Triglycerides; CE = cholesteryl ester.

to redistribute nonenzymically among other lipoproteins and plasma membranes. The action of LCAT on HDL subsequently converts this lipid to cholesteryl ester, which largely transfers to VLDL, IDL or LDL. The action of LCAT on HDL also leads to transfer of apolipoprotein E from subspecies of HDL to VLDL and IDL. It is proposed that the cholesteryl esters are then removed from the circulation either by hepatic mechanisms that depend on the recognition and removal of apolipoprotein E (fig. 2) or by the LDL pathway proposed by *Goldstein and Brown* [43]. Most cholesteryl esters would be removed by the apolipoprotein E pathway, some would be removed by the LDL pathway, and some would be removed by mechanisms that involve recognition of HDL. In the absence of LCAT, however, the pool of phosphatidylcholine and unesterified cholesterol that LCAT normally regulates would enlarge, binding sites for excess surface lipid would become saturated, and abnormal structures composed of this lipid would accumulate in the plasma.

Role of LCAT in Apolipoprotein Metabolism

As mentioned earlier, the content of C apolipoproteins in the VLDL of patients afflicted with familial LCAT deficiency is highly abnormal. The content of apolipoprotein C-I is high, whereas the contents of apolipoprotein C-II and C-III are low. Moreover, the content of each of the C apolipoproteins in the VLDL decreases when the patients' plasma is incubated with LCAT *in vitro*. Finally, model experiments using emulsions of triglyceride and phosphatidylcholine [94] have shown that addition of unesterified cholesterol to these emulsions affects the transfer of C apolipoproteins from normal HDL to the emulsions. It has been suggested [42], therefore, that the LCAT reaction is one of the factors that controls the distribution of apolipoprotein C between HDL and VLDL and contributes to the recycling phenomenon observed by others [53]. According to this view, the action of LCAT on nascent HDL increases their ability to bind and act as reservoirs for C apolipoproteins.

Role of the Reaction in Controlling the Content of Lipids in Cell Surface Membranes

The hypothesis that the LCAT reaction is part of a mechanism for transporting cholesterol from peripheral cells to the liver also was suggested several years ago [33]. Evidence at that time indicated that most peripheral cells can synthesize cholesterol, whereas degradation of cholesterol occurs mainly in the liver. Furthermore, as mentioned earlier, there was also evidence that the action of LCAT on HDL can displace unesterified cholesterol from erythrocyte membranes. It was suggested that the action of LCAT on HDL might generally displace unesterified cholesterol from peripheral cell membranes, and that HDL might transport this cholesterol to the liver. However, recent work in several laboratories has indicated that the interrelation between lipoproteins and peripheral cells is much more complex than implied by this hypothesis, so that it also must be modified.

The following considerations need to be taken into account. (1) If the above reasoning concerning chylomicron remnants is valid, then the content of lipids in peripheral cell surface membranes should vary somewhat throughout the day as a function of dietary fat intake. (2) To the extent that the erythrocyte lipids of patients with familial LCAT deficiency reflect this type of variation, not just unesterified cholesterol, but also phosphatidylcholine, phosphatidylethanolamine and sphingomyelin should be involved. (3) The work of *Brown and Goldstein* [43] and others [8] has clearly shown that peripheral cell requirements for cholesterol can be met not just by biosynthesis, but also by the receptor-mediated uptake of plasma lipoproteins. (4) If human peripheral cell membrane unesterified cholesterol is removed through the action of LCAT on HDL, much of the cholesteryl ester formed would be expected to transfer to VLDL, IDL or LDL, and this should make the cholesterol available for uptake into the cells by receptor-mediated processes, thus permitting recycling.

Thus, a modified hypothesis concerning the effect of the LCAT reaction on peripheral cell membranes can be formulated as follows: LCAT is one of the factors that controls the content of lipids on cell surface membranes to the extent that these lipids can equilibrate with HDL. Removal of unesterified cholesterol from these membranes through the action of LCAT on HDL can lead to a controlled recycling of cholesterol through the cells or to net removal of cholesterol from the cells. In the latter case, the cholesterol removed can either be transferred to other cells by mechanisms that depend on HDL receptors or it can (largely) be transferred to VLDL, IDL or LDL and to cells that have receptors for apolipoprotein B or apolipoprotein E. Although the liver is clearly the single most important pathway of cholesterol excretion, the mechanisms that mediate uptake of cholesterol by *human* livers remain to be determined.

Role of the Reaction in Other Aspects of Lipid Transport

Emphasis until now has been placed on potentially important roles of the LCAT reaction in humans. Studies in rats, however, strongly suggest that the reaction plays an entirely different role in this species. Studies by *Anderson and Dietschy* [8] and *Gwynne et al.* [48] have suggested that three specific rat organs have receptors for HDL: the adrenals, the ovaries and the testes. Since each of these organs synthesizes steroid hormones from cholesterol, it is apparent that HDL cholesteryl esters, formed by the LCAT reaction, may serve as an important source of cholesterol for steroid hormone synthesis. However, several observations suggest that the LCAT reaction may play an equally important role in the transport of arachidonic acid to these rat organs. The 'arachidonic acid transport pathway', proposed in figure 3, begins in the rat liver where arachidonic acid is synthesized from dietary linoleic acid. The arachidonic acid is then converted into HDL phosphatidylcholine, not by the Kennedy pathway of phosphatidylcholine synthesis, which mainly forms

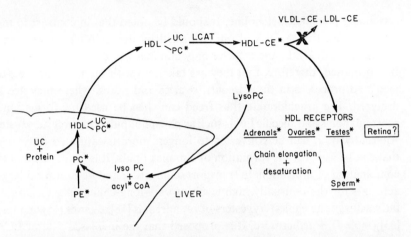

Fig. 3. Proposed pathway of arachidonic acid (*) transport in rats. Rat livers are known to make phosphatidylcholine that is rich in arachidonic acid by special pathways. This phosphatidylcholine is used to form lipoproteins that are secreted into the plasma. There action of LCAT on HDL leads to transfer of the arachidonic acid from phosphatidylcholine to cholesteryl ester. The HDL cholesteryl ester is then taken up by receptor-mediated pathways in the adrenals, ovaries, and testes. Arachidonate released in the cells of these organs is reesterified or used to form prostaglandins or longer-chain fatty acids. Critical features of the proposed pathway are: (1) formation by the liver of arachidonoyl phosphatidylcholine from lysophosphatidylcholine formed by LCAT; (2) control of LCAT activity by dietary essential fatty acids; (3) control of arachidonate and LCAT by estrogens; (4) lack of transfer of HDL cholesteryl esters to other lipoproteins; (5) evidence that adrenals, ovaries and testes contain receptors that are specific for HDL, and (6) high levels of arachidonic acid and its products in these three organs. CE = Cholesteryl ester; PC = phosphatidylcholine;   LysoPC = lysophosphatidylcholine;   PE = phosphatidylethanolamine.

phosphatidylcholines that contain oleate or linoleate, but by two alternate pathways that appear to be of special significance in rat liver [93]. The most important of these alternate pathways apparently involves an acylCoA:lysophosphatidylcholine acyltransferase that preferentially utilizes arachidonoyl CoA. The second alternate pathway appears to involve the sex-hormone-dependent methylation of hepatic phosphatidylethanolamine [57], which normally is rich in arachidonic acid. A subsequent step in the proposed arachidonic acid transport pathway occurs in the plasma, where LCAT transfers arachidonic acid from the C:2 position of HDL phosphatidylcholine to cholesteryl ester. The reaction also forms lysophosphatidylcholine which is rapidly removed by the liver [81] and reconverted into arachidonic-acid-containing phosphatidylcholine. It should be noted that in contrast to most other species, the cholesteryl esters formed by the LCAT reaction remain associated with HDL, and transfer only minimally to other lipoproteins [10]. It is proposed, therefore, that they are taken up with the HDL by receptor-mediated processes in the adrenals, ovaries and testes, where they are hydrolyzed. The arachidonate thus freed can then be used for formation of prostaglandins and specific lipids such as phospholipids and cholesteryl esters. Alternatively, it can be converted to longer, more unsaturated fatty acids, that can be used for the formation of complex lipids. It is suggested that this combination of arachidonate transport, elongation and desaturation largely accounts for the unusually high levels of very-long-chain, very unsaturated fatty acids in the cholesteryl esters of rat adrenals [18], ovaries [15] and testes [63] (table I). Furthermore, it is proposed that the increased rate of LCAT activity in rats fed diets that are deficient in essential fatty acids [87] reflects a control mechanism sensitive to the amount of essential fatty acid delivered to one or more of these organs. It may be that the content of arachidonic acid in target organs is critical since vitamin E deficiency in rats seems both to increase the formation of this acid from longer-chain fatty acids in the testes [16] and to decrease the rate of the LCAT reaction in the plasma [89]. In addition, since some effects of essential fatty acid deficiency can be reversed by injection of prostaglandin $F_{2\alpha}$, a derivative of arachidonic acid, it would clearly be of interest to determine whether this hormone affects LCAT activity in the rat. Another control mechanism that affects LCAT activity in rats and that may be related to the transport of essential fatty acids involves estrogens. Thus, it has been observed [1] that the plasma phospholipids and cholesteryl esters of female rats contain more arachidonate than do the corresponding plasma lipids of male rats and that LCAT activity also is higher in females than in males. The net effect of these sex-related differences might well be to increase transport of arachidonate to the ovaries. A final point that can be noted in relation to the proposed pathway for essential fatty acid transport in rats is that most of the vitamin E in rat plasma is known to be carried by HDL [12]. Thus, uptake of HDL by rat adrenals, ovaries and testes would be accompanied not only by uptake of essential fatty acids but also by a natural antioxidant.

Whereas the arachidonic acid transport pathway proposed above seems to be important in rats, essential fatty acid transport must be significantly different in humans. (1) The phosphatidylcholine of human HDL mainly contains linoleic acid instead of arachidonic acid. (2) As mentioned earlier, LCAT transfers less arachidonic acid to cholesterol than would be expected

Table I. Major cholesteryl esters of rat adrenals, testes and ovaries

Fatty acids	Adrenals [18] mol%	Testes [63] % weight	Ovaries [15] mol%
C16:0	17.3	20.9	6.7
C18:0	4.3	6.2	8.8
C18:1	21.1	10.9	12.3
C18:2	7.4	13.3	4.3
$\geq$C20:4	36.5	40.2	52.9

on the basis of the proportion of this fatty acid found in the C:2 position of phosphatidylcholine. (3) Most of the cholesteryl esters formed by the action of LCAT on HDL appear to transfer to VLDL, IDL or LDL. This would clearly make cholesteryl linoleate available to cells that have receptors for apolipoprotein B or apolipoprotein E as well as to cells that have receptors for HDL. Finally, the fatty acid composition of lipids in human adrenals, at least, is very different from that in rat adrenals in that the content of essential fatty acids or their derivatives is markedly reduced [18]. This suggests that the role of HDL cholesteryl esters in adrenal metabolism may differ in humans and rats.

Concluding Remarks: Unanswered Questions

The information reviewed in this chapter suggests that the biology of the LCAT reaction differs greatly in humans and rats. Consequently, the results obtained in studies of lipoprotein metabolism in the two species cannot be considered interchangeable and should be interpreted with caution. On the other hand, interspecies comparisons may prove very useful. For example, the fact that the turnover times of apolipoprotein A differ so considerably in humans and rats, whereas the rates of the LCAT reaction do not, supports the concept that most cholesteryl esters formed by the action of LCAT on human HDL transfer to other lipoproteins in plasma. Moreover, the possibility should be considered that this difference in cholesteryl ester metabolism is accompanied by a difference in the metabolism of apolipoprotein E. It is known already [88] that rat HDL contain considerable amounts of this apolipoprotein, whereas human HDL do not [17], and the possibility should be explored that the LCAT-induced transfer of apolipoprotein E from HDL to VLDL and IDL contributes to cholesterol ester metabolism in humans but not in rats. Species differences in essential fatty acid transport also merit evaluation. Since LCAT appears to play a role in the preferential transport of these fatty acids to rat adrenals and gonads, the possibility that it plays a similar role in humans should clearly be considered. Human sperm are extremely rich in very-long-chain, very unsaturated fatty acids of the α-linolenic family [16]. Since these acids are present in trace amounts, if at all, in human plasma cholesteryl esters, the possibility should be considered that they are transported to the testes by a pathway that is independent of both HDL and LCAT[3]. Finally, the fact that LCAT-derived cholesteryl esters transfer readily from HDL to other lipoproteins in humans raises the possibility that an important role of the LDL pathway in humans is to transport essential fatty acids to cells that contain receptors for apolipoprotein B.

Appendix

Calculation of Cholesteryl Ester Turnover Times

Concentrations of HDL cholesterol in human males and females are approximately 1,240 and 1,395 nmol/ml, respectively [6]. Assuming that approximately 74% of the HDL cholesterol is esterified, that the rates of the LCAT reaction in males and females are, respectively, 86.7 and 60.8 nmol/ml/h [58], and that LCAT acts only on HDL, then the turnover times of HDL cholesteryl ester in males and females can be estimated to be 10.5 and 17 h, respectively.

The concentration of HDL cholesteryl ester in rats seems to be about 900 nmol/ml [90]. If the rate of the LCAT reaction is on the order of 53 nmol/ml/h [83], and assuming the action of LCAT on HDL as indicated above, the turnover times of rat HDL cholesteryl ester can be estimated to be about 17 h.

The turnover time of cholesteryl esters in human lipoproteins that contain apolipoprotein B can also be estimated based on the rate of the LCAT reaction, if it is assumed that these cholesteryl esters are largely formed by the action of LCAT on HDL and subsequently transfer to VLDL, IDL or LDL. Such estimates must clearly be qualified since they neglect the contribution of intestinal ACAT activity and since the proportion of HDL cholesteryl ester that transfers to VLDL, IDL and LDL has not been clearly defined. [Incubation experiments with plasma from patients with familial LCAT deficiency, 68, suggest that the proportion of cholesteryl esters transferred may be as high as 90%.] With these qualifications, and assuming that the concentration of cholesteryl ester in VLDL + IDL + LDL is about 2,700 nmol/ml both in males and females [6], and that the proportion

---

[3] The role of the very-long-chain, very unsaturated fatty acids of sperm phospholipids remains an enigma. However, fatty acids of this type are known to be highly susceptible to oxidation and polymerization [6, 54]. Therefore, it is particularly intriguing that recent studies [27] of the response of sea urchin eggs to fertilization have shown that one effect of fertilization is to cause release of peroxidase from the egg into the sperm-filled medium.

of cholesteryl esters that transfer from HDL to these lipoproteins reflects the contents of cholesteryl esters in HDL and lipoproteins that contain apolipoprotein B in native plasma, then the turnover times for these esters can be estimated to be about 31 and 44 h for males and females, respectively (cholesteryl ester in lipoproteins that contain apolipoprotein B ÷ rate of the LCAT reaction × fraction of total plasma cholesteryl ester in lipoproteins that contain apolipoprotein B).

References

1 Aftergood, L. and Alfin-Slater, R.B.: Sex differences in plasma cholesterol-esterifying activity in rats. J. Lipid Res. 8: 126–130 (1967).
2 Akanuma, Y. and Glomset, J.A.: *In vitro* incorporation of cholesterol-^{14}C into very low density lipoprotein cholesteryl esters. J. Lipid Res. 9: 620–626 (1968).
3 Akanuma, Y. and Glomset, J.: A method for studying the interaction between lecithin:cholesterol acyltransferase and high density lipoproteins. Biochem. biophys. Res. Commun. 32: 639–643 (1968).
4 Akanuma, Y.; Kuzuya, T.; Hayashi, H.; Ide, T., and Kuzuya, N.: Positive correlation of serum lecithin:cholesterol acyltransferase activity with relative body weight. Eur. J. clin. Invest. 3: 136–141 (1973).
5 Akanuma, Y.; Mitchell, C.D., and Glomset, J.A.: unpublished results.
6 Albers, J.J.: Data from Northwest Lipid Research Clinic (personal commun.).
7 Albers, J.J.; Cabana, V.G., and Stahl, Y.D.B.: Purification and characterization of human plasma lecithin:cholesterol acyltransferase. Biochemistry, N.Y. 15: 1084–1087 (1976).
8 Andersen, J.M. and Dietschy, J.M.: Regulation of sterol synthesis in 16 tissues of rat. II. Role of rat and human high and low density plasma lipoproteins and of rat chylomicron remnants. J. biol. Chem. 252: 3652–3654 (1977).
9 Barter, P.J.: Origin of esterified cholesterol transported in the very low density lipoproteins of human plasma. J. Lipid Res. 15: 11–19 (1974).
10 Barter, P.G. and Lally, J.I.: The activity of an esterified cholesterol transferring factor in human and rat serum. Biochem. biophys. Acta (in press).
11 Bates, S.R. and Rothblatt, G.H.: Regulation of cellular sterol flux and synthesis by human serum lipoproteins. Biochim. biophys. Acta 360: 38–55 (1974).
12 Bjornson, L.K.; Gniewkowski, C., and Kayden, H.J.: Comparison of exchange of α-tocopherol and free cholesterol between rat plasma lipoproteins and erythrocytes. J. Lipid Res. 16: 39–53 (1975).
13 Blanchette-Mackie, E.J. and Scow, R.O.: Retention of lipolytic products in chylomicrons incubated with lipoprotein lipase; electron microscope study. J. Lipid Res. 17: 57–67 (1976).
14 Blum, C.B.; Levy, R.I.; Eisenberg, S.; Hall, M. III; Goebel, R.H., and Berman, M.: High density lipoprotein metabolism in man. J. clin. Invest. 60: 795–807 (1977).
15 Carney, J.A. and Walker, B.L.: Ovarian lipids from normal and essential fatty acid deficient rats during oestrus and dioestrus. Comp. Biochem. Physiol. 41B: 137–145 (1972).

16   Coniglio, J.: Gonadal tissue; in Snyder, Lipid metabolism in mammals, vol. 2, pp. 83–125 (Plenum Press, New York 1977).
17   Curry, M.D.; McConathy, W.J.; Alaupovic, P.; Leford, J.H., and Popovic, M.: Determination of human apolipoprotein E by electroimmunoassay. Biochim. biophys. Acta 439: 413–425 (1976).
18   Daily, R.E.; Swell, L.; Field, H., jr., and Treadwell, C.R.: Adrenal cholesterol ester fatty acid composition of different species. Proc. Soc. exp. Biol. Med. 105: 4–6 (1960).
19   Eder, H.A.: The lipoproteins of human serum. Am. J. Med. 23: 269–282 (1957).
20   Eisenberg, S. and Levy, R.I.: Lipoprotein metabolism. Adv. Lipid Res. 13: 2–80 (1975).
21   Eisenberg, S. and Rachmilewitz, D.: Metabolism of rat plasma very low density lipoproteins. I. Fate in circulation of the whole lipoprotein. Biochim. biophys. Acta 326: 378–405 (1973).
22   Faergeman, O. and Havel, R.J.: Metabolism of cholesterol esters of rat very low density lipoproteins. J. clin. Invest. 55: 1210–1218 (1975).
23   Faergeman, O.; Sata, T.; Kane, J.P., and Havel, R.J.: Metabolism of apoprotein B of plasma very low density lipoproteins in the rat. J. clin. Invest. 56: 1396–1403 (1975).
24   Fielding, C.J. and Fielding, P.E.: Purification and substrate specificity of lecithin: cholesterol acyltransferase from human plasma. FEBS Lett. 15: 355–358 (1971).
25   Fielding, C.J.; Shore, V.G., and Fielding, P.E.: Lecithin:cholesterol acyltransferase: effects of substrate composition upon enzyme activity. Biochim. biophys. Acta 270: 513–518 (1972).
26   Fielding, C.J.; Shore, V.G., and Fielding, P.E.: A protein cofactor of lecithin: cholesterol acyltransferase. Biochem. biophys. Res. Commun. 46: 1493–1498 (1972).
27   Foerder, C.A. and Shapiro, C.M.: Release of ovoperoxidase from sea urchin eggs hardens the fertilization membrane with tyrosine crosslinks. Proc. natn. Acad. Sci. USA 74: 4214–4218 (1977).
28   Gidez, L.I.; Roheim, D.S., and Eder, H.A.: Effect of diet on the cholesterol ester composition of liver and of plasma lipoproteins in the rat. J. Lipid Res. 6: 377–382 (1965).
29   Gidez, L.I.; Roheim, P.S., and Eder, H.A.: Turnover of cholesteryl esters of plasma lipoproteins in the rat. J. Lipid Res. 8: 7–15 (1967).
30   Gjone, E.; Norum, K.R., and Glomset, J.A.: Familial lecithin:cholesterol acyltransferase deficiency; in Stanbury, Wyngaarden and Fredrickson, The metabolic basis of inherited disease; 4th ed., pp. 589–603 (McGraw-Hill, New York 1978).
31   Glomset, J.A.: The mechanism of the plasma cholesterol esterification reaction: plasma fatty acid transferase. Biochim. biophys. Acta 65: 128–135 (1962).
32   Glomset, J.A.: Further studies of the mechanism of the plasma cholesterol esterification reaction. Biochim. biophys. Acta 70: 389–395 (1963).
33   Glomset, J.A.: The plasma lecithin:cholesterol acyltransferase reaction. J. Lipid Res. 9: 155–167 (1968).
34   Glomset, J.A.: Plasma lecithin:cholesterol acyltransferase; in Nelson, Blood lipids and lipoproteins: quantitation, composition and metabolism, pp. 745–787 (Wiley, New York 1972).
35   Glomset, J.A.: Lecithin:cholesterol acyltransferase; in Scanu, Biochemistry of atherosclerosis; 4th ed., pp. 589–603 (Dekker, New York, in press).
36   Glomset, J.A.; Janssen, E.T.; Kennedy, R., and Dobbins, J.: Role of plasma lecithin:

cholesterol acyltransferase in the metabolism of high density lipoproteins. J. Lipid Res. 7: 639–648 (1966).
37   Glomset, J.A.; Mitchell, C.; King, W.C., and Gjone, E.: unpublished.
38   Glomset, J.A.; Nichols, A.V.; Norum, K.R.; King, W.C., and Forte, T.: Plasma lipoproteins in familial lecithin:cholesterol acyltransferase deficiency. Further studies of very low and low density lipoprotein abnormalities. J. clin. Invest. 52: 1078–1092 (1973).
39   Glomset, J. and Norum, K.: The metabolic role of lecithin:cholesterol acyltransferase: perspectives from pathology. Adv. Lipid Res. 11: 1–65 (1973).
40   Glomset, J.A.; Norum, K.R., and King, W.C.: Plasma lipoproteins in familial lecithin:cholesterol acyltransferase deficiency: lipid composition and reactivity *in vitro*. J. clin. Invest. 49: 1827–1837 (1970).
41   Glomset, J.A.; Norum, K.R.; Nichols, A.V.; King, W.C.; Mitchell, C.D.; Applegate, K.R.; Gong, E.L., and Gjone, E.: Plasma lipoproteins in familial lecithin:cholesterol acyltransferase deficiency: effects of dietary manipulation. Scand. J. clin. Lab. Invest. 35: suppl. 142, pp. 3–30 (1975).
42   Glomset, J.A. and Verdery, R.B.: Role of LCAT in cholesterol metabolism; in Polonovski, Cholesterol metabolism and lipolytic enzymes, pp. 137–142 (Masson, New York 1977).
43   Goldstein, J.L. and Brown, M.S.: The low density lipoprotein pathway and its relation to atherosclerosis. A. Rev. Biochem. 46: 897–930 (1977).
44   Goodman, D.S.: The *in vivo* turnover of individual cholesterol esters in human plasma lipoproteins. J. clin. Invest. 43: 2026–2036 (1964).
45   Goodman, D.S.; Deykin, D., and Shiratori, T.: The formation of cholesterol esters with rat liver enzymes. J. biol. Chem. 239: 1335–1345 (1964).
46   Goodman, D.S. and Shiratori, T.: *In vivo* turnover of different cholesterol esters in rat liver and plasma. J. Lipid Res. 5: 578–586 (1964).
47   Goodman, D.S. and Shiratori, T.: Fatty acid composition of human plasma lipoprotein fractions. J. Lipid Res. 5: 307–313 (1964).
48   Gwynne, J.T.; Mahaffee, D.; Brewer, H.B., jr., and Nay, R.L.: Adrenal cholesterol uptake from plasma lipoproteins: regulation by corticotrophin. Proc. natn. Acad. Sci. USA 73: 4329–4333 (1976).
49   Hamilton, R.: Synthesis and secretion of plasma lipoproteins. Adv. exp. Med. Biol. 26: 7–24 (1972).
50   Hamilton, R.L.; Williams, M.C.; Fielding, C.J., and Havel, R.J.: Discoidal bilayer structure of nascent high density lipoproteins from perfused rat liver. J. clin. Invest. 58: 667–680 (1976).
51   Haugen, R. and Norum, K.R.: Coenzyme-A-dependent esterification of cholesterol in rat intestinal mucosa. Scand. J. Gastroenterol. 11: 615–621 (1976).
52   Havel, R.J.: Lipoproteins and lipid transport; in Kritchevsky, Paoletti and Holmes, Lipids, lipoproteins and drugs, pp. 37–60 (Plenum Press, New York 1975).
53   Havel, R.J.; Kane, J.P., and Kashyap, M.L.: Interchange of apolipoproteins between chylomicrons and high density lipoproteins during alimentary lipemia in man. J. clin. Invest. 52: 32–38 (1971).
54   Jones, R. and Mann, T.: Lipid peroxides in spermatozoa; formation, role of plasmalogen, and physiological significance. Proc. R. Soc. B 193: 317–333 (1976).

55 Kudchodkar, B.J. and Sodhi, H.S.: Plasma cholesteryl ester turnover in man: comparison of *in vivo* and *in vitro* methods. Clinina chim. Acta *68:* 187–194 (1976).
56 Langer, T.; Strober, W., and Levy, R.I.: The metabolism of low density lipoprotein in familial type II hyperlipoproteinemia. J. clin. Invest. *51:* 1528–1536 (1972).
57 Lyman, R.L.: Endocrine influence on the metabolism of polyunsaturated fatty acids; in Holman, Progress in the chemistry of fats and other lipids, vol. 9, pp. 195–230 Pergamon Press, Oxford 1968).
58 Marcel, Y.L. and Vezina, C.: A method for the determination of the initial rate of reaction of lecithin:cholesterol acyltransferase in human plasma. Biochim. biophys. Acta *306:* 497–504 (1973).
59 Marcel, Y.L. and Vezina, C.: Lecithin:cholesterol acyltransferase in human plasma. J. biol. Chem. *218:* 8254–8259 (1973).
60 Mckenzie, I.F.C. and Nestel, P.J.: Studies on the turnover of triglyceride and esterified cholesterol in subjects with nephrotic syndrome. J. clin. Invest. *47:* 1685–1695 (1968).
61 Mead, J. and Fulco, A.: The unsaturated and polyunsaturated fatty acids in health and disease (Thomas, Springfield 1976).
62 Murphy, J.: Erythrocyte metabolism. III. Relationship of energy metabolism and serum factors to the osmotic fragility following incubation. J. Lab. clin. Med. *60:* 86–109 (1962).
63 Nakamura, M.; Jensen, B., and Privett, O.S.: Effect of hypophysectomy on the fatty acids and lipid classes of rat testes. Endocrinology *82:* 137–142 (1968).
64 Nichols, A.V. and Gong, E.L.: Use of sonicated dispersions of mixtures of cholesterol with lecithin as substrates for lecithin:cholesterol acyltransferase. Biochim. biophys. Acta *231:* 175–184 (1971).
65 Nichols, A.V. and Smith, L.: Effect of very low density lipoproteins on lipid transfer in incubated serum. J. Lipid Res. *6:* 206–210 (1965).
66 Nilsson, A. and Zilversmit, D.B.: Release of phagocytosed cholesterol by liver macrophages and spleen cells. Biochim. biophys. Acta *260:* 479–491 (1972).
67 Norum, K. and Gjone, E.: The influence of plasma from patients with familial plasma lecithin:cholesterol acyltransferase deficiency on the lipid pattern of erythrocytes. Scand. J. clin. Lab. Invest. *22:* 94–98 (1968).
68 Norum, K.R.; Glomset, J.A.; Nichols, A.V.; Forte, T.; Albers, J.J.; King, W.C.; Mitchell, C.D.; Applegate, K.R.; Gong, E.L.; Cabana, V., and Gjone, E.: Plasma lipoproteins in familial lecithin:cholesterol acyltransferase deficiency: effects of incubation with lecithin:cholesterol acyltransferase *in vitro*. Scand. J. clin. Lab. Invest. *35:* suppl. 142, pp. 31–55 (1975).
69 Olivecrona, T.: personal commun.
70 Quarfordt, S.H. and Goodman, D.S.: Metabolism of doubly labeled chylomicron cholesteryl esters in the rat. J. Lipid Res. *8:* 264–273 (1967).
71 Raz, A.: Action of lecithin-cholesterol acyltransferase on sonicated dispersions of lecithin and cholesterol and on lecithin-cholesterol-protein complexes. Biochim. biophys. Acta *239:* 458–468 (1973).
72 Redgrave, T.G.: Cholesterol feeding alters the metabolism of thoracic duct lymph lipoprotein cholesterol in rabbits but not in rats. Biochem. J. *136:* 109–113 (1973).
73 Rehnborg, C.S. and Nichols, A.V.: The fate of cholesteryl esters in human serum incubated *in vitro* at 38°. Biochim. biophys. Acta *84:* 596–603 (1964).

74 Roheim, P.S.; Hirsch, H.; Edelstein, D., and Rachmilewitz, D.: Metabolism of iodinated high density lipoprotein subunits in the rat. III. Comparison of the removal rate of different subunits from the circulation. Biochim. biophys. Acta *278:* 517–529 (1972).

75 Roheim, D.S.; Haft, D.E.; Gidez, L.I.; White, A., and Eder, H.A.: Plasma lipoprotein metabolism in perfused rat livers. II. Transfer of free and eserified cholesterol into the plasma. J. clin. Invest. *42:* 1277–1285 (1963).

76 Rose, H.G.: High density lipoproteins: substrates and products of plasma lecithin: cholesterol acyltransferase; in Day and Levy, High density lipoproteins (Plenum Press, New York, in press).

77 Rose, H.G. and Juliano, J.: Regulation of plasma lecithin:cholesterol acyltransferase in man. I. Increased activity in primary hypertriglyceridemia. J. Lab. clin. Med. *88:* 29–43 (1976).

78 Rose, H.G. and Juliano, J.: Regulation of plasma lecithin:cholesterol acyltransferase. II. Activation during alimentary lipemia. J. Lab. clin. Med. *89:* 525–532 (1977).

79 Schumacher, V.N. and Adams, G.H.: Circulating lipoproteins. A. Rev. Biochem. *38:* 113–136 (1969).

80 Sigurdsson, G.; Nicoll, A., and Lewis, B.: Conversion of very low density lipoprotein to low density lipoprotein. A metabolic study of apolipoprotein B kinetics in human subjects. J. clin. Invest. *56:* 1481–1490 (1975).

81 Stein, Y. and Stein, O.: Metabolism of labeled lysolecithin, lysophosphatidyl ethanolamine, and lecithin in the rat. Biochim. biophys. Acta *116:* 95–107 (1966).

82 Stein, Y.; Stein, O., and Goren, R.: Metabolism and metabolic role of serum high density lipoproteins; in Gotto, jr., Miller and Oliver, High density lipoproteins and atherosclerosis, pp. 37–49 (Elsevier/North Holland, Amsterdam 1978).

83 Stokke, K.T.: Cholesteryl ester metabolism in liver and blood plasma of various animal species. Atherosclerosis *19:* 393–406 (1974).

84 Stokke, K.T. and Norum, K.R.: Determination of lecithin:cholesterol acyltransfer in human blood plasma. Scand. J. clin. Lab. Invest. *27:* 21–27 (1971).

85 Subbaiah, P.V. and Bagdade, J.D.: Demonstration of enzymic conversion of lysolecithin to lecithin in normal human plasma. Life Sci. (in press, 1978).

86 Sugano, M. and Portman, O.W.: Fatty acid specificities and rates of cholesterol esterification *in vivo* and *in vitro*. Archs Biochem. Biophys. *107:* 341–351 (1964).

87 Sugano, M. and Portman, O.W.: Essential fatty acid deficiency and cholesterol esterification activity of plasma and liver *in vitro* and *in vivo*. Archs Biochem. Biophys. *109:* 302–315 (1965).

88 Swaney, J.B.; Braithwaite, F., and Eder, H.A.: Characterization of the apolipoproteins of rat plasma lipoproteins. Biochemistry, N.Y. *16:* 271–278 (1977).

89 Takatori, T. and Privett, O.S.: Studies of serum lecithin:cholesterol acyl transferase activity in rat: Effect of vitamin E deficiency, oxidized dietary fat, or intravenous administration of ozonides or hydroperoxides. Lipids *9:* 1018–1023 (1974).

90 Ugazio, C. and Lombardi, B.: Serum lipoproteins in rats with ethionine-induced fatty liver. Lab. Invest. *14:* 711–719 (1965).

91 Utermann, G.; Menzel, H.J., and Langer, R.H.: On the polypeptide composition of an abnormal high density lipoprotein (LP-E) occurring in lecithin:cholesterol acyltransferase deficient plasma. FEBS Lett. *45:* 29–32 (1974).

92 Utermann, G.; Menzel, H.J.; Langer, R.H., and Dieker, P.: Lipoproteins in familial

lecithin:cholesterol acyltransferase deficiency. II. Further studies on the abnormal high density lipoproteins. Humangenetik *27:* 185–187 (1975).
93 Van Golde, L.M.G. and van den Bergh, S.G.: in Snyder, Lipid metabolism in mammals, vol.1, pp.35–116 (Plenum Press, New York 1977).
94 Warth, D.C.; Erkelens, D.W., and Glomset, J.A.: Effect of unesterified cholesterol on apolipoprotein transfer from high density lipoproteins to an artificial fat emulsion. Clin. Res. *25:* A130 (1977).
95 Zilversmit, D.B.; Hughes, L.B., and Balmer, J.: Stimulation of cholesterol ester exchange by lipoprotein-free rabbit plasma. Biochim. biophys. Acta *409:* 393–398 (1975).

J.A. Glomset, MD, Howard Hughes Medical Institute Laboratory,
Departments of Medicine and Biochemistry, and Regional Primate Research Center,
University of Washington, Seattle, WA 98195 (USA)

Intravascular Metabolism of Lipoproteins

Kinetic Analysis of Turnover Data

Mones Berman[1]

Laboratory of Theoretical Biology, DCBD, NCI
National Institutes of Health, Bethesda, Md.

Introduction

The levels of lipoprotein particles in plasma or in various organs are established by their rates of entry or formation and their turnover or residence times. The former is determined by the availability of precursors, by synthesis processes and by conversion from other substrates. The latter are determined by metabolic or catabolic processes and the local environmental factors within which the processes take place. To determine the rates of formation or the turnover of the particles, steady-state data alone are usually inadequate and kinetic studies are necessary. Kinetic data are basically measurements of transient responses of systems to inputs or perturbations. The analysis and interpretation of the response curves require the *proposal* of a mathematical or mechanistic model, the *calculation of the model response* to a simulated input or perturbation, the *fitting* of the calculated response to the experimental observations by adjustment of the model parameters and, finally, the *identification* of the model parameters with the parameters of the physiological system. The model thus serves to transform the experimental data to physiological parameters of interest. The nature and complexity of the model and the analysis of the data depend on the extent of the data and on the physiological parameters being considered. Thus, a variety of models may be utilized for the same system, depending on the objective.

For a complex system, the greater the diversity of data – the more complex the model is of necessity to explain all the data. If one limits oneself to the

---

[1] I wish to thank Drs. David Foster and Loren Zech for their useful comments about the paper.

Table I. Nomenclature

$a_i(t)$	Specific activity of compartment i. Ratio of tracer to tracee
$f_i(t)$	Tracer amount in compartment i
$f_{ij}(t)$	Tracer amount in compartment i due to input to $u_j(t)$ into compartment j only
A_j	Magnitude (intercept) of jth exponential component: $A_j e^{-\alpha_j t}$
α_j	Exponential coefficient in $e^{-\alpha_j t}$
$w(t)$	Response of linear, time invariant system to unit impulse
$r(t)$	Response of system to an arbitrary input $u(t)$
$c_p(t)$	Concentration of tracer in plasma
C_p	Steady-state concentration of tracee in plasma
M_p	Steady-state mass in plasma
M_{ij}	Steady-state mass in compartment i due to a constant input U_j to compartment j
M_i	Steady-state mass in compartment i
M_T	Total steady-state mass of system
V_P	Plasma volume
V_i	Plasma equivalent volume (PEV) of compartment i: tracee mass in compartment i divided by plasma tracee concentration, $V_i = M_i/C_p$
$u_i(t)$	Tracer input rate into compartment i from outside the system
U_i	Steady-state input rate of new tracee material into compartment i from outside the system
U_T	Total steady-state input rate of new tracee material into the system $U_T = \Sigma U_i$
L_{ij}	Rate constant: fraction of material in compartment j that flows to compartment i per unit time. When $i=0$ (L_{oi}) the flow is irreversible to the 'outside'. L_{jj} is the sum of all rate constants out of compartment j: $L_{jj} = L_{oj} + L_{1j} + L_{2j} + \ldots$
IC_i	Initial conditions in compartment i
PCC	Plasma-containing compartment: plasma plus all other spaces in exchange with it that are too rapid to resolve and therefore treated as a single compartment
PR_i	Production rate of *new* tracee material (first entry) for compartment i. This includes direct entry from outside the system (U_i), and first time entries contributed through other compartments
IDR_i	Irreversible disposal rate: the rate of loss material from compartment i by all pathways never to return. At steady state, this equals the production rate, PR_i
σ_{ij}	Fraction of particles leaving compartment j that reach compartment i
$\tilde{t}_{ij}$	Expected residence time in compartment i for particles that enter compartment j
$\tilde{t}_i$	$= \tilde{t}_{ii}$
FCR_{ij}	$= 1/\tilde{t}_{ij}$
FCR_i	Fractional catabolic rate for particle in compartment i: fraction of particles lost per unit time irreversibly from compartment i via all exit pathways
FCR_p^*	Fractional catabolic rate for tracer injected into plasma: fraction of injected tracer lost per unit time irreversibly from plasma via all exit pathways
FCR_p^c	Fractional catabolic rate for plasma tracee: fraction of material in plasma lost per unit time irreversibly from plasma via all exit pathways
MCR_p	Metabolic clearance rate: plasma equivalent volume lost per unit time irreversibly from the plasma via all exit pathways. $MCR_p = FCR_p \cdot V_p$
p	Used as subscript to designate plasma
~	Used to designate integration from $t=0$ to $t \to \infty$: $\tilde{x} = \int_0^\infty x(t)\,dt$
*	Used as superscript to designate tracer
.	Used to designate derivative with respect to time: $\dot{x} = dx/dt$

identification of only a few parameters of the system, it may be possible to get by with less extensive data or with simpler models. When simpler models are used, however, not all the information in the data is usually extracted, and additional assumptions about the system and its behavior outside the range of observations are implied. The validity of these assumptions can only be tested if a fuller knowledge of the system is available, which is usually not the case. Herein lies the 'fly in the ointment'. For any given set of data and assumptions, a model is consistent if it fits the data. Whether the identification of physiological parameters is truly valid, however, depends on the extent to which the assumptions have been validated. This can explain why different investigators using assumptions and data bases which are not comparable frequently arrive at different conclusions about the same system. On the other hand, a model that describes a system more fully (1) integrates and extracts more information from the data, (2) provides a greater understanding of the system and the limitations of the data, (3) permits the examination of the validity of simpler modeling approaches, and (4) makes it possible to incorporate known physical and physiological constraints into the model thereby restricting the permissible classes of models and adding to the confidence of the derived parameter values.

In studying lipoprotein kinetics, certain measures are frequently derived: production or entry rates, fractional catabolic rates, and plasma and extraplasma masses or spaces. I shall describe some simplified mathematical models commonly used to derive this information. I will then show more detailed models for the system and examine possible sources of errors in the use of the simpler models. It should be realized that the more detailed models presented here still do not fully identify the true system so that a complete validation cannot be made.

The notation employed in this paper is defined in table I.

Lipoprotein Kinetics

I shall restrict my discussion here to the kinetics of four classes of plasma lipoproteins based on density separation: very-low-density lipoproteins (VLDL; $d < 1.006$), intermediate-density lipoproteins (IDL; $1.006 < d < 1.019$), low-density lipoproteins (LDL; $1.019 < d < 1.063$) and high-density lipoproteins (HDL; $1.063 < d < 1.21$). Chylomicrons are not dealt with here, but some of the considerations presented apply to them too. For each of the classes of particles, there are a number of distinct moieties

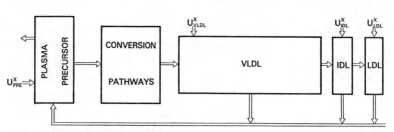

Fig. 1. Schematic diagram for the kinetics of labels in the various moieties (apoproteins, TG, etc.) in lipoprotein particles, including their precursors.

that may be labeled: apoproteins (A-I, A-II, B, C-I, etc.), triglycerides (TG), phospholipids (PL), cholesterol (FC) and cholesterol esters (CE). Each moiety has its own kinetic patterns. The kinetics of a particle as a whole are governed by the kinetics of the individual moieties. The kinetics are further complicated by the nature of the label. For example, ^{14}C and ^{3}H, even if attached to the same molecule, do not necessarily behave identically [63], and 'foreign' labels (e.g. ^{131}I) may interfere with metabolic pathways or not label all molecules uniformly.

A general diagram showing the kinetic stages for individual lipoprotein moieties is shown in figure 1. The elementary plasma precursors in this diagram are free fatty acids (FFA), amino acids, glycerol, mevalonic acid, acetate, etc. Either the precursors or the particles can be labeled, and the latter can be labeled exogenously or endogenously. As can be seen from the diagram, a labeled moiety introduced anywhere in the system can be propagated around the loop. Under certain conditions, the label can also cross over from one moiety to another (e.g., TG to PL, or FC to CE, etc.).

Kinetically, the system shown schematically can be further broken down into a number of subsystems. For example, plasma glycerol is part of a precursor subsystem for TG. Similarly, each of the moieties in the classes of particles – apolipoprotein B (apoB), apolipoprotein A-I (apoA-I), TG etc. – form separate subsystems.

A subsystem may be studied by labeling it directly or by labeling its precursors, and observing its responses and/or the responses of its products. Some of the information derived from these various procedures is equivalent. By carrying out several experimental procedures jointly, more extensive information is derived about the system and a certain degree of validation is obtained from the redundant information.

Computational Methods

There is a considerable body of theories on systems and kinetic analysis as well as on modeling and data analysis [2, 5–12, 14, 16, 24–26, 28, 30, 31, 37, 38, 41, 43, 47, 48, 50, 51, 56, 59–61, 65]. This needs to be applied when full extraction of information from complex or diverse data (in terms detailed biological mechanisms) is desired. When only limited information is of interest (e.g., distribution spaces, total exchangeable mass, fractional catabolic rate, residence time, etc.), simpler mathematical approaches can sometimes be used.

Most kinetic analyses involve tracer data obtained after a tracer is introduced into the system by a particular path. From such data, it is desired to estimate steady-state tracee (substance being traced) levels and transports. It is obvious, yet subtle, that tracer kinetics reflect only the tracer, and that to extend the tracer results to the tracee, additional information or assumptions are required. It is subtle because, as will be shown later, this is frequently not fully recognized in the analysis of kinetic data. The results of the tracer analysis apply directly to the tracee when the latter follows identical input and exchange pathways, but some information about the tracee can also be extracted when this condition is not fully satisfied. The extent of the information that can be extracted and the validity of implied assumptions in doing this can best be evaluated when the interrelationship between the tracer model and the tracee system is fully known.

Most of the tracer studies on the lipoprotein system are carried out under steady-state conditions of the tracee. Under these conditions, the kinetics of the tracer are linear[2] and the parameters of the tracer system do not change with time [9]. The differential equations describing such a system are said to be linear, constant-coefficient. Under these conditions, certain mathematical simplifications are possible to aid in the analyses, e.g., fitting data to sums of exponential, using areas under response curves, etc. These simplifications do not necessarily carry over to tracers in systems where the tracee is not in a steady state, or to transient kinetics of the tracee. In these cases, it is simplest to model the linear or nonlinear differential equations that describe the system directly and fit the parameters of the differential equations to the data. Computer programs to do this are available [15, 33, 39; see *Groth*, 22, for review].

---

[2] Definition of linear system: given that for an input $u_1(t)$ the response is $r_1(t)$ and for an input $u_2(t)$ the response is $r_2(t)$, a system is said to be linear if *any* arbitrary combination of inputs, $k_1u_1(t)+k_2u_2(t)$ yields the same combination of outputs: $k_1r_1(t)+k_2r_2(t)$.

Compartmental Analysis

In this method of analysis [30, 51], the system is simulated by a finite number of states, called compartments, with exchange of material between them. Material enters and leaves the system via these compartments. A schematic for a simple two-compartment model is shown in figure 2. The L_{ij} are rate constants expressing the fraction of material in compartment j (both tracer and tracee) that enters compartment i per unit time. The subscript o in L_{oj} is used to designate losses to the outside. $f_i(t)$ is the amount of tracer in compartment i and M_i is the amount of tracee in it. $u_i(t)$ is defined as the rate of entry of new tracer material into compartment i directly from outside the system and U_i is the corresponding amount for the tracee. PR_i is the *total* rate of entry of *new* tracee material (first-time entries) into compartment i, including entries via other compartments.

The relations between amounts of tracer in the compartments and the rate constants are given by the following set of ordinary differential equations:

$$\frac{df_i}{dt} = \sum_j L_{ij} f_j(t) - \sum_j L_{ji} f_i(t) + u_i(t), \qquad (1)$$

where the L_{ij} are constant or time-dependent.

An arbitrary response, $r_k(t)$, is usually defined by some linear combination of compartmental contents:

$$r_k(t) = \sum_i S_{ki} f_i(t), \qquad (2)$$

where S_{ki} is the fraction of $f_i(t)$ in the response $r_k(t)$.

For L_{ij} constant, the steady-state tracee levels, M_i, and input rates, U_i, are related through

$$U_i = \sum_j L_{ji} M_i - \sum_j L_{ij} M_j. \qquad (3)$$

There are various procedures for generating compartmental models from tracer response curves [8–12, 14, 16, 24–26, 37, 41, 47, 56] and for the calculation of their rate constants. For steady-state tracee systems, the tracer curves are frequently fitted to sums of exponentials and these are then used to calculate the L_{ij} of the model [14, 37, 41]. Such a procedure, however, is limited to relatively simple models. Alternatively, special conditions may permit the calculation of a particular rate constant. For example, the amount of substance

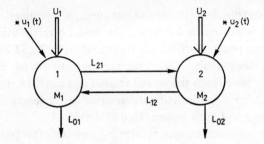

Fig. 2. General two-compartment model. The notation is explained in table I. For this model,

$$FCR_1 = L_{01} + L_{21}\frac{L_{02}}{L_{12} + L_{02}},$$

$$PR_1 = U_1 + \sigma_{12}U_2 = U_1 + \frac{L_{12}}{L_{12} + L_{02}}U_2 = FCR_1 \cdot M_1,$$

$$M_2 = (L_{21}M_1 + U_2)/(L_{12} + L_{02}),$$

$$U_T = U_1 + U_2 = L_{01}M_1 + L_{02}M_2.$$

Given that the response in compartment 1, $f_1(t)$, to a unit bolus injection into that compartment is a sum of two exponentials

$$(f_1(t) = A_1 e^{-\alpha_1 t} + A_2 e^{-\alpha_2 t},$$

where $A_1 + A_2 = 1$) the following can be uniquely defined:

$$L_{11} = L_{21} + L_{01} = A_1\alpha_1 + A_2\alpha_2$$
$$L_{22} = L_{12} + L_{02} = A_1\alpha_2 + A_2\alpha_1$$
$$L_{12}L_{21} = L_{11}L_{22} - \alpha_1\alpha_2$$

L_{12}, L_{21}, L_{01} and L_{02} cannot be uniquely determined without some additional data or assumption. If it is assumed that $L_{02}=0$, all parameters become uniquely determined:

$$L_{01} = 1/(A_1/\alpha_1 + A_2/\alpha_2)$$
$$L_{21} = L_{11} - L_{01}$$
$$L_{12} = (L_{11}L_{22} - \alpha_1\alpha_2)/(L_{11} - L_{01}).$$

The residence times can be defined in terms of the L_{ij} as follows. Let $D = L_{11}L_{22} - L_{12}L_{21}$. Hence, $\bar{t}_{11} = L_{22}/D$, $\bar{t}_{12} = L_{21}/D$, $\bar{t}_{2,1} = L_{12}/D$, $\bar{t}_{22} = L_{11}/D$.

excreted into urine from plasma over a time interval t_1–t_2 may be calculated from the equation:

$$\text{Urine} = k\int_{t_1}^{t_2} f_p(t)\,dt,$$

or variations of this equation [4], where $f_p(t)$ is the amount of substance in plasma and k is a rate constant. A more general and practical method, applicable to small and large systems, both linear and nonlinear, is the numerical solution of the differential equations that describe the model directly using computers. General computer programs to do this are available [15, 22, 33, 39]. This approach, although somewhat 'brute force', permits the testing of models against all available data at once (both tracer and tracee) and can also yield statistical measures of confidence for derived parameter values or functions. It is the simplest, most direct and most general tool available.

Given a set of L_{ij}, all steady-state masses, residence times and other parameters can be calculated[3] for any set of constant inputs U_i. Conversely, given steady-state masses and sites of entry of new material, calculation of the rates of entry are possible (equation 3). When there is ambiguity about the model constants and/or the sites of entry of new material, unique determination of production rates of new material is not possible. At best, these values can be predicted within a specified range. The following example illustrates this.

Consider a tracer input, $u_i(t)$, into compartment $\underline{i}$ only and measurements in the same compartment, $f_i(t)$. A number of models can be generated that are compatible with such data. The following four special cases indicate the effects on the prediction of the steady-state tracee quantities. In this example $\underline{i}$ represents the measured compartment and $\underline{j}$ represents compartments other than i.

(A) All tracee input and irreversible losses occur into and from the measured compartment i ($U_j = 0$, $L_{oj} = 0$ for all j).
(B) Tracee input is into compartment $\underline{i}$ but irreversible losses occur from other compartments ($U_j = 0$, $L_{oi} = 0$, $L_{oj} \neq 0$).
(C) All tracee input occurs in compartments outside the measured compartment, but all irreversible losses occur from the measured compartment only ($U_i = 0$, $U_j \neq 0$, $L_{oj} = 0$ for $j \neq i$).
(D) All tracee input and losses occur from non-sampled compartment ($U_i = 0$, $L_{oi} = 0$, $U_j \neq 0$, $L_{oj} \neq 0$).

For these cases, the results for fractional catabolic rate for compartment $\underline{i}$, FCR_i, production rate for compartment $\underline{i}$, PR_i, total production rate, U_T, and total system mass, M_T, are given in table II.

[3] In matrix notation, if L is the matrix of the L_{ij}, the elements of the inverse matrix L^{-1} are the steady-state solutions for a set of unit input vectors given by the columns of the unit matrix. These inverse elements are also the residence times of particles in the compartments for different entry sites [31]. The fractional catabolic rates are the reciprocal of these residence times.

Table II. Comparison of calculated results for various input and output pathways in compartmental model with tracer input into and sampling from compartment i

	Case A $U_j = 0$ $L_{oj} = 0$	Case B $U_j = 0$ $L_{oi} = 0$	Case C $U_i = 0$ $L_{oj} = 0$	Case D $U_i = 0$ $L_{oi} = 0$
FCR_i	$= L_{oi}$	= case A	= case A	= case A
PR_i	$= FCR_i M_i$	= case A	= case A	= case A
U_T	$= PR_i$	= case A	= case A	$\geq$ case A
M_T	$= \dfrac{\int_0^\infty t f_i(t)}{\int_0^\infty f_i(t)}$	$>$ case A	$\geq$ case A	$\geq$ case A

Mass of compartment i, M_i, is assumed known. In all cases $j \neq i$, and values of U and L are generally nonzero unless otherwise specified.

Residence Time

A measure frequently used in kinetics is residence time, which has been discussed extensively in the literature [5, 31, 38, 48, 51, 59, 61]. Because residence time relates to fractional catabolic rate (FCR), steady-state masses and areas under tracer curves, and follows a simple algebra, we present a short discussion of it to provide a greater insight into these relations.

We define *residence time* ($\bar{t}_{ij}$) to mean the expected time that a particle in or entering into compartment j will spend in compartment i. Numerically, this quantity equals the steady-state amount in compartment i contributed by a constant unit input rate into compartment j.

The expected residence time for a particle in any block of compartments, say X, when first introduced into compartment j is the sum of the individual residence times of the compartments contained in X:

$$\bar{t}_{Xj} = \sum_X \bar{t}_{ij}. \tag{4}$$

A particle may have entry sites into several compartments with probability $U'_j = U_j/U_T$ (fraction of total) for compartment j. Its expected residence time in compartment i is the weighted mean

$$\bar{t}_i = \sum_j U'_j \bar{t}_{ij}. \tag{5}$$

For a block of compartments i designated by X having a distribution of inputs to compartments j,

$$\bar{t}_X = \sum_i \sum_j U'_j \bar{t}_{ij}. \tag{6}$$

where i is summed over the compartments contained in X.

Since numerically $\bar{t}_{ij}$ equals the mass in compartment $\underline{i}$ due to a unit infusion of tracee into compartment j, the residence times can be expressed in terms of steady-state quantities:

$$\bar{t}_{ij} = \frac{M_{ij}}{U_j}, \tag{7}$$

where M_{ij} is the mass in compartment $\underline{i}$ contributed by the input into compartment j only (U_j). The total mass in compartment $\underline{i}$ is

$$M_i \equiv \sum_j M_{ij} = \sum_j U_j \bar{t}_{ij} \tag{8}$$

and for the total mass of the system, M_T,

$$M_T \equiv \sum_i \sum_j M_{ij} = \sum_i \sum_j U_j \bar{t}_{ij}. \tag{9}$$

When a tracer is administered to compartment $\underline{j}$ at a rate $u_j(t)$ and the amount of tracer, $f_{ij}(t)$, due to this input is observed in compartment $\underline{i}$, then

$$\frac{\int_0^\infty u_j(t)\,dt}{\int_0^\infty f_{ij}(t)\,dt} = \frac{U_j}{M_{ij}}. \tag{10}$$

Using equation (7),

$$\bar{t}_{ij} = \frac{M_{ij}}{U_j} = \frac{\int_0^\infty f_{ij}(t)\,dt}{\int_0^\infty u_j(t)\,dt}. \tag{11}$$

Thus, equation (11) relates the residence time with tracer data and with steady-state tracee quantities.

It should be pointed out that the residence time is also uniquely determined by the rate constants of a system. For a compartmental model, this can be expressed as the inverse of the matrix of rate constants [31]. For a two-compartment model, this is given in the caption of figure 2.

Fractional Catabolic Rate

Conventionally, FCR for a compartment $\underline{i}$, FCR_i, is defined as the reciprocal of the residence time, $\bar{t}_i$, in that compartment for material entering it. It is convenient to generalize this definition as follows:

$$\text{FCR}_{ij} \equiv \frac{1}{\bar{t}_{ij}} = \frac{U_j}{M_{ij}} = \frac{\int_0^\infty u_j(t)\,dt}{\int_0^\infty f_{ij}(t)\,dt} \tag{12}$$

and

$$\text{FCR}_X = \frac{1}{\bar{t}_X}.$$

When $i=j$, i.e. the observed (response) compartment is also the tracee input compartment,

$$\text{FCR}_{ii} \equiv \text{FCR}_i \equiv \frac{1}{\bar{t}_i} = \frac{U_i}{M_{ii}} = \frac{\int_0^\infty u_i(t)\,dt}{\int_0^\infty f_{ii}(t)\,dt} \tag{13}$$

When all inputs are into compartment $\underline{i}$ only, namely, $U_T = U_i$, $U_T(t) = U_i(t)$, then $M_i = M_{ii}$, $f_{ii}(t) = f_i(t)$, and

$$\text{FCR}_i = \frac{U_i}{M_i} = \frac{\int_0^\infty u_i(t)\,dt}{\int_0^\infty f_i(t)\,dt} \equiv \frac{\tilde{u}_i}{\tilde{f}_i}, \tag{14}$$

where $\tilde{u}_i$ is the total tracer input and $\tilde{f}_i$ is the area under the curve for compartment i. Equation (14) corresponds to the more common use of FCR, in particular, when $\underline{i}$ is the plasma compartment.

Production rate for compartment $\underline{i}$, PR_i, is defined as the first-entry input rate into compartment i necessary to account for the mass M_i [24]. It contains direct input from the outside into compartment $\underline{i}$, U_i, and first-entry contributions from inputs into other sites in the system:

$$\text{PR}_i = \frac{M_i}{M_{ii}} U_i = U_i + \sigma_{ij} U_j, \tag{15a}$$

where σ_{ij} is the fraction of material that enters compartment j from the outside (U_j) that also serves as first entries for compartment $\underline{i}$.

Using equation (13), we can rewrite equation (15a) as

$$\text{PR}_i = \frac{M_i}{\bar{t}_i} = \text{FCR}_i \cdot M_i. \tag{15b}$$

This is a widely used relation to calculate the production rate of 'new' material for compartment $\underline{i}$. Note that because the same material may serve as 'new' material for more than one compartment, U_T is in general *not* equal to $\sum_i \text{PR}_i$.

Metabolic clearance rate, MCR_i, is defined [62] as

$$\mathrm{MCR}_i = \mathrm{FCR}_i \cdot V_i = \frac{\tilde{u}_i}{\tilde{f}_i} \cdot V_i = \frac{\tilde{u}_i}{\tilde{c}_i}. \qquad (16)$$

Simplified Calculations and Special Cases

Plasma is frequently a site of entry of new material as well as a site of sampling. Material in plasma, however, may rapidly equilibrate with surrounding tissues forming kinetically a compartment having a space different than that of plasma. It may be difficult to establish from available data what the space of distribution of this plasma containing compartment is, in which case it is more practical to deal with plasma data in units of concentration or specific activity.

Let the subscript p represent plasma or a plasma-containing compartment, and let a fraction b of new tracer entering the system pass (first entry) through this compartment. Let $\tilde{u}_T$ be the total amount of tracer administered. Then using equation (14), $\mathrm{FCR}_p = b\tilde{u}_T/\tilde{f}_p = b\tilde{u}_T/\tilde{a}_p M_p$ where $\tilde{a}_p$ is the area under the specific activity curve. Combining this with equation (15b) for the tracee, we get

$$\mathrm{FCR}_p = \frac{b\tilde{u}_T}{\tilde{a}_p M_p} = \frac{\mathrm{PR}_p}{M_p}. \qquad (17a)$$

Hence,

$$\mathrm{PR}_p = b\frac{\tilde{u}_T}{\tilde{a}_p}. \qquad (17b)$$

This equation states that if the amount of new tracer entering the plasma-containing compartment is known, and the plasma specific activity curve is given, then the total production rate of new *tracee* entering this compartment, PR_p, is uniquely determined.

Similarly, let B be the fraction of the total tracee input into the system that must pass through plasma, i.e. $\mathrm{PR}_p = B \cdot U_T$. Equation (17b) may then be restated as

$$U_T = \frac{b}{B}\frac{\tilde{u}_T}{\tilde{a}_p}. \qquad (17c)$$

Equation (17c) states that given the total amount of tracer administered, $\tilde{u}_T$, and the area under the plasma specific activity curve, $\tilde{a}_p$, the total production rate for the system, U_T, can be determined if the ratio b/B is known. In particular when the tracee follows identical input pathway to tracer b/B = 1 and the well-known relation of *Bergner* [5] follows:

$$U_T = \frac{\tilde{u}_T}{\tilde{a}_p}. \tag{17d}$$

For the case when *all* losses from the system occur from plasma or the plasma-containing compartment, the average time that particles introduced into plasma spend in the total system, $\bar{t}_s$, is given by [31, 38, 48, 59]:

$$\bar{t}_s = \frac{\int_0^\infty t f_p(t)\, dt}{\int_0^\infty f_p(t)\, dt} = \frac{\widetilde{tf_p}}{\tilde{f}_p} = \frac{\widetilde{ta_p}}{\tilde{a}_p}. \tag{18}$$

For $f_p(t)$ expressed as a sum of exponentials [59]

$$\bar{t}_s = \frac{\sum_i \dfrac{A_i}{\alpha_i^2}}{\sum_i \dfrac{A_i}{\alpha_i}}. \tag{19}$$

The mass of the total system, M_T, in this case, given that all new material U_T enters the plasma-containing compartment, is

$$M_T = \bar{t}_s \cdot U_T = \bar{t}_s \cdot \frac{\tilde{u}_T}{\tilde{a}_p} = \frac{\widetilde{ta_p}}{\tilde{a}_p^2} \cdot \tilde{u}_T. \tag{20}$$

For response functions f(t) represented by sums of exponentials,

$$f(t) = \sum_i A_i e^{-\alpha_i t},$$

the following relations apply:

$$\int_0^\infty f(t)\, dt = \sum_i \frac{A_i}{\alpha_i}[(1 - e^{-\alpha_i t})]_0^\infty = \sum_i \frac{A_i}{\alpha_i} \tag{21a}$$

$$\int_0^\infty t f(t)\, dt = \sum_i \frac{A_i}{\alpha_i^2}[1 - e^{-\alpha_i t}(1 + \alpha_i t)]_0^\infty = \sum_i \frac{A_i}{\alpha_i^2} \tag{21b}$$

$$\left.\frac{df(t)}{dt}\right|_{t=0} = \sum_i -A_i \alpha_i e^{-\alpha_i t}\bigg|_{t=0} = -\sum_i A_i \alpha_i \tag{21c}$$

$$f(o) = \sum_i A_i. \tag{21d}$$

Cascade Process

It has been necessary in studies of VLDL-apoB (VLDL-B) and VLDL-TG kinetics to introduce a cascade process, probably to represent various stages

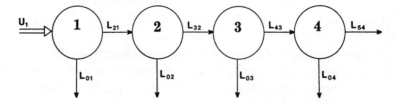

Fig. 3. A cascade model. Particles enter compartment 1. Each successive state represents an increasing density band as particles are delipidated. Particles may also be lost irreversibly through L_{oi}. If β represents the fraction of mass passing on from one state to the next ($\beta = L_{21}/(L_{21}+L_{01})$), then the fraction that reaches IDL in this four-state model is β^4.

of delipidation of the particles [13, 44, 45, 64]. Such a process has been approximated by a number of states or compartments in tandem as shown in figure 3. The four states in the figure represent VLDL particles or their moieties (apoB, TG, CH, etc.) in various stages of delipidation, starting with newly formed particles, represented by compartment 1. The L_{oi} represent irreversible loss of particles or partial loss of a moiety from a particle in state i such as loss of TG or apoC. For simplicity, it is convenient to represent the VLDL particles as a cascade of states having similar turnover rates, namely $L_{11} = L_{22} = L_{33} = L_{44} = L_{ii}$, where $L_{ii} = L_{i+1,i} + L_{oi}$.

Let $\bar{t}$ be the residence time for a moiety in the entire cascade and let $\bar{t}_i = 1/L_{ii}$ be its residence time in compartment i. Let $\beta = L_{i+1,i}/L_{ii}$ be the fraction of material transferred from one state in the cascade to the next, and $(1-\beta)$ the fraction that is lost irreversibly. Hence, for a particle entering compartment 1, the residence time in the cascade having n states is

$$\bar{t} = \bar{t}_1 + \beta \bar{t}_2 + \beta^2 \bar{t}_3 + \beta^3 \bar{t}_4 + \cdots + \beta^{n-1}\bar{t}_n = \bar{t}_i(1+\beta+\beta^2+\beta^3+\cdots+\beta^{n-1}) = \bar{t}_i \frac{1-\beta^n}{1-\beta}. \quad (22)$$

When $\beta = 1$, i.e. irreversible loss of material from the cascade occurs only from the last compartment, as is frequently the case for VLDL-B, $\bar{t} \equiv \bar{t}^o = n\bar{t}_i$. $\bar{t}^o$ may be viewed as the time it takes a VLDL particle to delipidate to an extent that it leaves the VLDL density range. On the other hand, when, say, $\beta = 0.75$ and $n = 4$ (which approximates the case for VLDL-TG), $\bar{t} = 2.73\,\bar{t}_1$. Hence, the residence time (reciprocal FCR) for VLDL-B is about 50% greater than for VLDL-TG when $n = 4$. This ratio varies with the initial size of particles and with the number of states in the cascade. In particular, when a single compartment is used, the residence times for TG and apoB are the same.

Table III. Relative residence times, $\bar{t}/\bar{t}^o$, for cascade models having different number of states, n, and varying fractions, k, of material pass through the entire cascade

Number of states, n	Fraction passed through entire cascade $k = \beta^n$					
	1	0.80	0.41	0.30	0.10	0.0625
1	1.00	1.00	1.00	1.00	1.00	1.00
2	1.00	0.95	0.82	0.77	0.66	0.62
3	1.00	0.93	0.76	0.71	0.56	0.52
4	1.00	0.92	0.74	0.67	0.51	0.47
8	1.00	0.91	0.69	0.63	0.45	0.40
16	1.00	0.90	0.68	0.60	0.42	0.37
∞	1.00	0.90	0.66	0.58	0.39	0.34

$\bar{t}^o$ is the residence time of a particle in the cascade when no losses occur during passage, except for the final exit loss, i.e., when $\beta = 1$.

A comparison of residence times for various values of n and β is presented in table III.

The preceding calculates residence times for particles entering the first state of the cascade. The residence time for a distribution of entry sites is the weighted mean of the individual residence times. For four states, letting U'_i be the fraction of the total input entering state $\underline{i}$,

$$\bar{t} = U'_1 \frac{\bar{t}^o}{4}(1 + \beta + \beta^2 + \beta^3) + U'_2 \frac{\bar{t}^o}{4}(1 + \beta + \beta^2) + U'_3 \frac{\bar{t}^o}{4}(1 + \beta) + U'_4 \frac{\bar{t}^o}{4} \quad (23)$$

$$= \frac{\bar{t}^o}{4}\left[1 + (U'_1 + U'_2 + U'_3)\beta + (U'_1 + U'_2)\beta^2 + U'_1 \beta^3\right].$$

A cascade implies a certain physiological process. It suggests that material lost from a particle occurs in steps, and that the *particle* stays in the cascade until the combined losses shift its density to a new range. Since each moiety on a particle may lose differing fractions at each step of a cascade, the residence times for different moieties on the same particle may vary, and the residence time of the particle as a whole is a function of the individual losses. When the cascade degenerates to a single compartment the implication is that the *entire particle* is lost from the cascade in a single step. Hence, for this condition the residence times for all moieties including the particle as a whole are the same.

Convolution – Deconvolution

A first order system is considered time-invariant (or linear, constant parameter), if the kinetics are first-order with time-invariant rate constants. A tracer in a steady-state tracee environment follows such kinetic [9].

For a linear time-invariant system, given a response, w(t), to a unit impulse, the response, r(t), to any arbitrary input u(t) can be calculated from the convolution integral [47, 50, 59]

$$r(t) = \int_0^t u(x) \cdot w(t-x)\, dx, \tag{24}$$

where x is a dummy variable of integration. For example, if the response to a *unit* impulse is $w(t) = e^{-kt}$, then the response r(t) to an input $u(t) = Ae^{-\alpha t}$ is

$$r(t) = \int_0^t A e^{-\alpha x} \cdot e^{-k(t-x)}\, dx = A\, \frac{e^{-kt} - e^{-\alpha t}}{\alpha - k}. \tag{25}$$

Convolution is commutative so that the roles of w(t) and u(t) can be interchanged:

$$r(t) = \int_0^t u(t-x) w(x)\, dx. \tag{26}$$

The inverse process to convolution is called deconvolution, i.e., given a response r(t) and an input, u(t), calculate the unit response, w(t); or, given r(t) and w(t), calculate u(t). Except for simple functions, convolution and deconvolution are best calculated using computers. A convenient method for use with the SAAM program [15] is described elsewhere [6].

Computer Modeling

When a model is nonlinear or complex the analytical solution of the responses to various inputs or perturbations become difficult or impossible and one has to resort to numerical solutions. Computers are ideal for this and many modeling programs have been developed for this purpose [see *Groth*, 22, for review]. There are additional benefits that arise from computer modeling. Complementary information in diverse forms (parameter relations, initial conditions constraints, discrete observations, etc.) can readily be incorporated, thus extending the data base for testing a model. Furthermore, statistical tests can be utilized to yield confidence limits for derived parameter values and using the covariance matrix derived in connection with least-squares fitting of the data, error estimates can be derived for any function of the parameters. The computer of course yields results rapidly and once a program is fully tested, the computations are essentially error-free. Comparison of

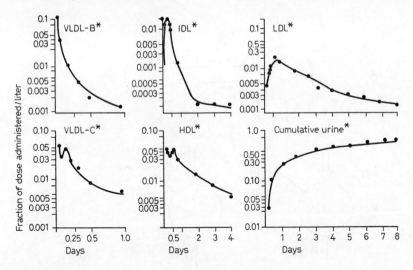

Fig. 4. Typical apoB and apoC tracer data for a normal subject after unit bolus injection of VLDL-B* and VLDL-C* [14]. ● = observed values; —— = predicted values.

various models and the simulation of potential experiments add further to the power of modeling with the aid of the computer.

Lipoprotein Subsystems

VLDL-B

The simplest model used to explain VLDL-B kinetics in plasma is a single compartment. Such a model is consistent with studies of exogenously labeled VLDL-B over a period of about 4–8 h after injection. Studies over longer periods of time [19, 45], however, show a second exponential emerging at about 8 h or so (fig. 4). This suggests that at least two compartments are involved and, depending on the way these compartments interrelate, determines the qualitative and quantitative interpretations for the kinetics. The following 2-compartment models have been considered.

(A) A homogeneous population of labeled VLDL-B in plasma in exchange with an extraplasma compartment (fig. 5A). This model is challenged by the plasma IDL-apoB (IDL-B) data derived from labeled VLDL-B [13, 44]. Since, in this model, labeled IDL-B is a direct product of the homogeneous plasma compartment of VLDL-B*, the entire VLDL-B* curve should serve as the precursor for IDL-B*. In many cases studied, however, this does not

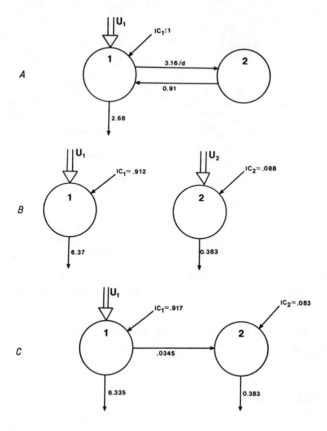

Fig. 5. Three two-compartment models compatible with observed plasma tracer curves after the injection of exogenously labeled VLDL-B. Typical plasma curve, normalized to unity at t = 0, is given by $f_p(t) = A_1 e^{-\alpha_1 t} + A_2 e^{-\alpha_2 t} = 0.912\, e^{-6.37 t} + 0.088\, e^{-0.383 t}$. $V_p = 1/c_p(0)$. *A* In this model, there is one plasma population in exchange with an extraplasma compartment. $IC_1 = 1$; $IC_2 = 0$; $L_{01} = 1/\bar{f}_p$; $L_{11} = \dot{f}_p(0)$; $L_{21} = L_{11} - L_{01}$; $L_{22} = L_{12} = \alpha_1 + \alpha_2 - L_{11}$; $FCR_p^c = FCR_p^* = L_{01}$; $M_1 = M_p$; $PR_p = U_p = L_{01} M_p$; $M_2 = L_{21} M_1 / L_{12}$, and $M_T = M_1 + M_2 = M_p(1 + L_{21}/L_{12})$. *B* In this model, there are two populations of particles and it is assumed that each population is initially labeled in proportion to its apoB mass. Hence, $IC_1/IC_2 = M_1/M_2$; $L_{01} = \alpha_1$; $L_{02} = \alpha_2$; $IC_1 = A_1$; $IC_2 = A_2$; $M_p = M_1 + M_2$; $U_1/U_2 = A_1 \alpha_1 / A_2 \alpha_2$; $U_1/U_T = A_1 \alpha_1 / (A_1 \alpha_1 + A_2 \alpha_2)$; $U_2/U_T = A_2 \alpha_2 / (A_1 \alpha_1 + A_2 \alpha_2)$, and $FCR_p^c = A_1 \alpha_1 + A_2 \alpha_2$. $FCR_p^* = 1/(A_1/\alpha_1 + A_2/\alpha_2)$. *C* In this two-population model, it is assumed that each population is initially labeled in proportion to its apoB mass, hence, $IC_1/IC_2 = M_1/M_2$; $L_{11} = \alpha_1$; $L_{02} = \alpha_2$; $IC_2 = A_2(\alpha_1 - \alpha_2)/\alpha_1$; $IC_1 = 1 - IC_2$; $L_{21} = \alpha_2 \cdot IC_2/IC_1$, and $L_{01} = \alpha_1 - L_{21}$; for $U_2 = 0$ $FCR_p^c = L_{11}/(1 + L_{21}/L_{22})$.

seem to be the case. If one satisfies the early IDL-B* data, the calculated slow component for IDL-B* is larger than observed. This suggests that the slow component in VLDL-B* is kinetically a differently labeled population from the fast component.

(B) Two populations of VLDL-B particles in plasma, each produced *de novo* directly (fig. 5B). This model is not favored in light of results obtained in cross-over experiments [13, 44]. When VLDL-B from a type IV subject is exogenously labeled and injected into a patient with a type III hyperlipoproteinemia, the VLDL-B* slow component observed is greater than that produced when injected into a normal subject. This suggests that some of the *injected* label must be converted to the slow component. Additional data, however, are required to validate this more fully.

For this model, exogenous labeling produces a tracer input distribution proportional to the masses of the two moieties and not their input rates. Hence, the FCR for the tracer is not the same as for the tracee, as seen in table IV.

(C) Two populations of VLDL-B particles in plasma with the rapidly turning-over population serving as a precursor for the slow one (fig. 5C). This hypothesis is favored here since it has been found to be compatible with data obtained from various experiments.

Of special note here is the fact that particles entering the slowly turning-over population contribute to a larger steady-state mass than those entering the fast population, because of the increased residence time. Hence, when VLDL-B is exogenously labeled, the slow-population particles, if initially labeled in proportion to their mass, show a greater slow component in the plasma curve compared to their relative production rate. Endogenously generated labeled VLDL-B produces a labeled slow component in direct proportion to the production rate, and is therefore much smaller than that obtained from the exogenous label.

Results of the analysis using model C and comparison to models A and B are presented in table IV.

Careful examination of the VLDL-B kinetic curves over the first few hours suggests that a simple exponential decay does not always describe them adequately [44, 45, 64]. The curves show some flattening at the early points. This is further accompanied by a delay in the appearance of labeled apoB in IDL. These kinetic features can be generated by a cascade process discussed earlier and as shown in figure 3, which is interpreted as a stepwise delipidation process of VLDL particles. Such a process has been experimentally shown to exist [3, 4, 40, 52, 58].

Table IV. Parameter values for plasma VLDL-B and VLDL-TG subsystems for several models

Parameter	ApoB						TG, five-compartment model
	all models	two-compartment model			five-compartment model	integral method	
		A	B	C			
C_p (measured), mg/ml	0.0456						0.0616
V_p (estimated), ml	2,870						2,870
M_p, mg	131		$M_1 = 119.5$ $M_2 = 11.5$				177
M_T, mg		585	131	131	131	585	177
FCR_p^*, d^{-1}	2.68						4.56
FCR_p^c, d^{-1}		2.68	5.84	5.84	3.66	2.68	4.56
PR_p, mg/d		351				351	806
U_p, mg/d		351	765 $\begin{cases} U_1 = 761 \\ U_2 = 4.4 \end{cases}$	765	479	351	806
MCR, l/d		7.7	16.8	16.8	10.5	7.7	13.1

VLDL-B tracer (exogenously labeled with iodine) administered to plasma as unit bolus. Plasma curve can be described by $f_p(t) = 0.912\ e^{-6.37t} + 0.088\ e^{-0.383t}$, where t is in days. $\bar{t}_p = 0.373\ d = 1/2.68$; $\dot{f}_p(0) = 5.84/d$, and $\widetilde{tf}_p = 0.622\ d^2$. TG results are based on direct fit of data to five-compartment model based on glycerol kinetics [64]. Similar model for FFA given by Shames et al. [49].

Although a cascade process can be represented by a large number of states, each having its own mean time $\bar{t}_i$, a minimal number consistent with the resolution of the data is chosen [13, 44, 45]. If the number of stages chosen is too small, the flatness is not fully reproduced. On the other hand, if too many stages are introduced, the transitions from flatness to curvature are too sharp. Four stages appeared adequate to represent this process. Again, because of the limited resolution, each state was assumed to have the same mean time $\bar{t}_i = \bar{t}_j = 1/L_{ii} = 1/L_{jj}$.

The parameter that characterizes the cascade, in addition to the number of states, is the mean residence time, $\bar{t}$, of a particle in the entire cascade. As the number of states increases, this value becomes relatively independent of the number of states chosen (see table III).

The residence time of VLDL-B in the cascade of states reflects the time it takes for a VLDL particle to delipidate to the IDL density level, and on other loss processes. The former depends on the rate of delipidation (lipase activity, apoC-II, etc.) and the initial size of the particle. It has been suggested [23, 58] that as particles delipidate and become smaller, their rate of delipidation decreases. Available kinetic data do not seem to require this [64], but cannot exclude it either.

Let us consider a simple cascade of four states (fig. 3) with $\bar{t}_i = 1/L_{ii}$ as the residence time per state and with $\beta = 1$, i.e., loss of VLDL-B from the final state only. As discussed in the section on 'Computational Methods', the total residence time of a particle entering the cascade by way of the first state is $\bar{t} = \Sigma \bar{t}_i$. For all $L_{ii} = 1/\bar{t}_i$ equal to each other, $\bar{t} = 4\bar{t}_i = 4/L_{ii}$; $FCR_p = L_{ii}/4$; $U_1 = FCR_p \cdot M_p$, and $M_1 = M_2 = M_3 = M_4 = (1/4) M_p$. When VLDL-B is exogenously tagged and the label for each compartment is proportional to its mass, i.e., one quarter of the label in each, the average residence time for the injected tracer, $\bar{t}^*$, is (equation 23)

$$\bar{t}^* = (1/4)(4\bar{t}_i) + (1/4)(3\bar{t}_i) + (1/4)(2\bar{t}_i) + (1/4)(\bar{t}_i) = 2.5\bar{t}_i = 0.625\bar{t}.$$

Hence, the calculated $FCR_p^* = 1/\bar{t}^*$ for the tracer is 1.6 times greater than for the tracee entering compartment 1 only. There is some evidence [58] that particles are produced in a spectrum of sizes. Hence, there are inputs U_1, U_2, ..., etc. The expected residence time and FCR for tracee particles in this case will still not correspond to that calculated using an exogenous tracer, since the latter labels in proportion to the plasma masses and not the input rates U_i.

We shall now consider the cascade process in the context of model C (fig. 5), in which compartment 1 is replaced by four cascade states (compartments 11–14, fig. 6), and each state feeds the same fraction $(1-\beta)$ of its particles to a slowly turning-over compartment 2.

The residence time in plasma of VLDL-B entering compartment 11 in this model is the sum of the residence times in the cascade states and in compartment 2. Let $\bar{t}_c$ be the residence time for the cascade and let $\bar{t}_2 = 1/L_{o2}$ be the residence time for compartment 2. Then the total residence time is

$$\bar{t} = \bar{t}_c + q_2 \bar{t}_2, \tag{27}$$

where q_2 is the fraction of the material entering the cascade that is diverted to compartment 2. If all the material that does not go on to IDL-B is diverted to compartment 2, then $q_2 = 1 - \beta^4$, and from equations (22) and (27).

$$\bar{t} = \frac{1}{FCR} = \frac{\bar{t}^o}{4} \frac{1-\beta^4}{1-\beta} + (1-\beta^4)\bar{t}_2 = \bar{t}_c \left[1 + (1-\beta^4)\frac{\bar{t}_2}{\bar{t}_c} \right]. \tag{28}$$

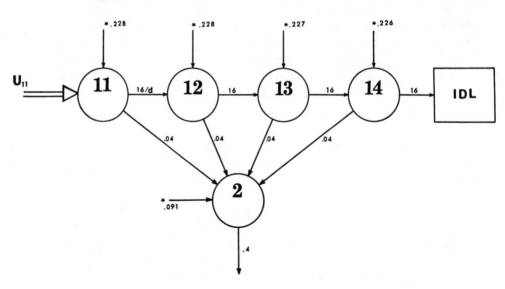

Fig. 6. VLDL-B model containing delipidation cascade (compartments 11, 12, 13 and 14) and a second population of particles (compartment 2). In this model, all of the newly synthesized VLDL-B particles enter compartment 11 and about 1% of these are diverted to compartment 2, which has a residence time about 10 times longer than the cascade. The initial conditions reflect the distribution of label resulting from the exogenous labeling, assuming equal efficiencies in labeling for each state or population. The values given are approximations for a normal subject.

For normals $\bar{t}_2/\bar{t}_c \approx 10$ and $1-\beta^4 \approx 0.01$. Hence $\bar{t}=1.1\,\bar{t}_c$. In patients with type III disorder, however, $1-\beta^4 \approx 0.19$ and $\bar{t}=2.9\,\bar{t}_c$.

VLDL-TG

The models described for VLDL-B also apply to VLDL-TG, except for the additional loss of TG in the delipidation process.

In most studies, plasma VLDL-TG is endogenously labeled by using plasma FFA or glycerol precursors [20, 27, 32, 46, 49]. A typical [64] plasma labeled VLDL-TG curve derived from a plasma glycerol precursor is shown in figure 7. The kinetics of the precursors and their conversion to plasma VLDL-TG have been studied by a number of investigators, using various methods of analysis. A discussion of some of these has been published by *Shames et al.* [49], and the most recent model is presented by *Zech et al.* [64].

It was commonly assumed in the past on the basis of short studies that labeled plasma FFA and glycerol precursors, because they rapidly disappear

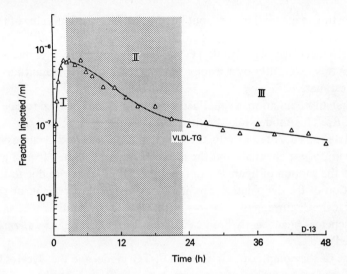

Fig. 7. Plasma VLDL-TG tracer curve following a unit bolus injection of glycerol into plasma [64].

from plasma, quickly label the plasma VLDL-TG pool, and that subsequently the VLDL-TG label disappears monoexponentially from the plasma. Longer studies, however, reveal that after about 8–12 h, the plasma VLDL-TG curve develops a second, much slower exponential component. Several serious complications seem to arise as a result of this. The slow component can arise from at least three sources [49]. The first is a population of slowly turning over VLDL-TG particles, as observed for VLDL-B. The second is a slowly turning over precursor pool somewhere in the synthesis pathway which continues to supply labeled VLDL-TG for a long time. The third is an exchange of plasma VLDL-TG with other moieties or non-plasma spaces. These complications, together with the need for a delipidation chain make the interpretation of the VLDL-TG kinetics more complex [64].

If the tracer enters plasma VLDL-TG by the same route as the tracee (likely, but not necessarily), then the FCR for VLDL-TG can be calculated using the integral-equation approach

$$FCR_p = \frac{\tilde{u}_p}{\tilde{f}_p},$$

where $\tilde{f}_p$ is the integral of the plasma curve and $\tilde{u}_p$ is the total input into plasma VLDL-TG. Unfortunately, this equation, cannot be readily applied since it is difficult to estimate the total tracer input into VLDL-TG, $\tilde{u}_p$. The value of the initial peak of the VLDL-TG curve could be an approximation of the total input from the fast glycerol to VLDL-TG pathway. A slow pathway, howev is can contribute significantly over a long time interval, and its magnitudeer, difficult to estimate.

To a first approximation, a simplified model (fig. 8) may be used to calculate the FCR_p. This model, may be fit directly to the data and all the rate constants determined. Alternatively, the labeled plasma VLDL-TG curve can be fit to a sum of exponentials and the rate constants calculated from that, as shown in the caption of figure 8. Some complications can arise due to the approximation of the delipidation cascade by a single compartment, or the presence of a slowly turning over moiety of VLDL-TG. Evaluation of several simplified schemes has recently been carried out by *Ye, Grundy and Berman* (unpublished).

Because of the complexities in the VLDL-TG model and the glycerol to VLDL-TG conversion pathways, an estimate for FCR_p is best obtained by modeling the entire system, including the glycerol or FFA precursors [49, 64], with the aid of a computer (fig. 9). This approach also provides confidence estimates for the derived parameter values.

Available data support the notion that all the triglyceride FFA derive from plasma. Labeled plasma FFA therefore parallel the tracee as precursors. Although plasma glycerol accounts for only 10% of the total tracee VLDL-TG glycerol, it may still be used to calculate the FCR_p of plasma VLDL-TG, since all other sources of glycerol probably enter the VLDL-TG subsystem via the same plasma compartments.

The FCR values for apoB and for TG can be compared using equation (22) ($L_{ii} = 1/\bar{t}_i$). For a normal subject, using the four-state cascade, $L_{ii} = 16/d$, $L_{01} = 0$ ($\beta = 1$) for apoB, and $L_{ii} = 16/d$, $L_{01} = 4.0/d$ ($\beta = 0.75$) for TG:

$$\bar{t}_p^B = \frac{4}{L_{ii}},$$

$$\bar{t}_p^{TG} = \frac{2.73}{L_{ii}}.$$

Hence,

$$\bar{t}^{TG} = 0.683\,\bar{t}^B.$$

The FCR ($=1/\bar{t}$), therefore, for VLDL-TG is about 50% higher than for VLDL-B. If a single-compartment model were correct, the FCR for VLDL-TG and VLDL-B would be identical.

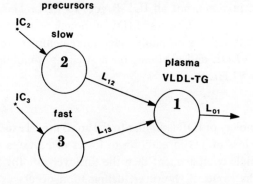

Fig. 8. Simplified model for the calculation of VLDL-TG kinetic parameters. Compartment 1 is plasma VLDL-TG. Compartments 2 and 3 represent a fast and slow conversion pathway, respectively, from plasma glycerol to plasma VLDL-TG. IC_2 and IC_3 are initial conditions that need to be determined from the data fitting, together with the parameters L_{12} and L_{13}. Given that the VLDL-TG curve is fitted to a sum of three exponentials, $f_p(t) = A_1 e^{-\alpha_1 t} + A_2 e^{-\alpha_2 t} + A_3 e^{-\alpha_3 t}$, with $\alpha_1 > \alpha_2 > \alpha_3$, then $FCR_p^c = L_{01} \approx A_1/(\bar{f}_p - A_1/\alpha_1 - A_3/\alpha_3)$; $L_{12} = \alpha_1$; $L_{13} = \alpha_3$.

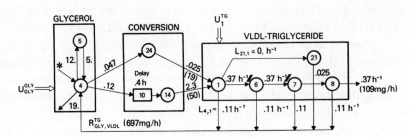

Fig. 9. VLDL-TG model with its synthesis pathways and glycerol subsystem [64]. The glycerol subsystem is that derived by *Malmendier et al.* [36]. The TG 'conversion' subsystem contains a fast pathway (compartments 10 and 14) and a slow pathway (compartment 24). The plasma VLDL-TG subsystem is comparable to that for VLDL-B. Compartment 21 is the slow decaying moiety and compartments 1, 6, 7 and 8 constitute the delipidation cascade. Values shown are approximations for normal subjects ($L_{21,1}$ assumed negligible). FCR_p^c for the plasma VLDL-TG subsystem can be calculated from equation 28, and the value is given in table VI.

IDL-B

IDL-B is usually modeled as a single compartment, intermediate between VLDL-B and LDL-B. Previous studies [13] suggest that not all the IDL-B necessarily comes from VLDL-B. In some individuals, IDL-B may be synthesized directly and similarly, in some, not all IDL-B goes to LDL. This is especially true in abnormal states. The kinetics of IDL, although represented by a simple one-compartment model is probably much more complex since it contains remnants from VLDL and chylomicron as well as β-mobility particles similar to those for VLDL-B.

IDL-TG

The model for the TG moiety of IDL is similar to the apoB model except that there is an irreversible loss of TG in addition to losses of particles. If IDL truly behaves like a single compartment then the turnover rate for its apoB and TG should be the same. If, however, delipidation occurs as a cascade process then the apoB and TG turnover rates will not be the same, as was pointed out for VLDL. Present data are inadequate to permit a distinction between these processes.

LDL-B

The kinetics of LDL particles have been studied most extensively by labeling their apoB. It is thought that apoB is an integral part of the LDL particle and that its metabolism is identical to that of the particle. This is not true for other moieties in LDL, such as LDL-TG and LDL-FC.

A typical set of kinetic data obtained for LDL after a single injection of iodinated LDL-B is shown in figure 10. The plasma data can be approximated by a sum of two exponentials, and the kinetics are usually explained [34] by a two-compartment model (model A; fig. 11). Compartment 1 is considered as plasma and compartment 2 in exchange with it is thought to be mostly liver [57] although this has not been fully validated. The degradation site of LDL-B is mostly from the plasma compartment as suggested from more extensive kinetic analyses of iodide excretions into urine [21]. The results for the two-compartment analysis are given in table V.

When LDL-B is labeled with iodide, the activity collected in urine can also be used as a direct measure of LDL degradation rate [17, 63]. An assumption usually made in connection with this is that the rate of accumulation of labeled iodide in urine is proportional to the level of labeled LDL-B in plasma, with the proportionality constant being the FCR. Actually, when iodide is released from LDL-B it equilibrates first with an iodide space and is therefore delayed before it appears in urine. The iodide space is about 9 times greater than plasma (35 vs. 4% body weight) and turns over at a rate of about 2/d (when the thyroid is blocked), with most of it cleared by the kidneys (fig. 11). In using urine to plasma activity ratio as a measure of LDL degradation, a

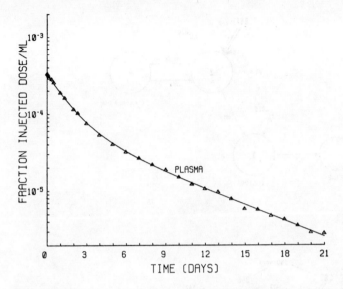

Fig. 10. Plasma tracer curve after a bolus injection of iodinated LDL-B into plasma.

correction for iodide delay is necessary. A shift of the urine data by half a day towards zero time can, to a first approximation, correct for the iodide delay. Neglect of the delay due to the iodide pool tends to overestimate the FCR value, while neglect of losses of label to nonurine pathways underestimates the value.

Kinetic analyses of the conversion of a VLDL-B and IDL-B to LDL-B suggest that most of the newly produced LDL enters the plasma compartment first. Hence, U_4 is assumed to be zero and $U_2 = U_T = PR_2$ (fig. 11A). The steady-state solutions for the tracee based on this assumption and on the model are given in table V.

Careful analysis of labeled LDL-B kinetics, however, using plasma and urine data simultaneously suggests that even with an iodide compartment the model is not fully consistent with the data. First, two exponentials are not quite adequate to fit the plasma curve. Secondly, no combination of degradations from the two exchange compartments of the LDL-B model, together with an iodide delay compartment, adequately fits the urine data [21]. To eliminate these inconsistencies, it was necessary to consider two moieties of labeled LDL-B (fig. 11B) and three compartments. One moiety requires a single compartment (compartment 1) and is totally confined to plasma. The

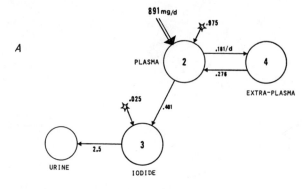

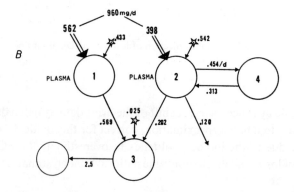

Fig. 11. LDL-B kinetic models. *A* Conventional two-compartment model used for the analysis of labeled plasma LDL-B data. Compartment 2 is plasma and compartment 4 is an extraplasma compartment. Compartment 3 is total body iodide (about 35% body weight). Given that the plasma LDL-B response is a sum of two exponentials, $f_p(t) = A_1 e^{-\alpha_1 t} + A_2 e^{-\alpha_2 t}$, $A_1 + A_2 = 1$, we have: $L_{32} = 1/(A_1/\alpha_1 + A_2/\alpha_2)$; $L_{22} = A_1\alpha_1 + A_2\alpha_2$; $L_{42} = L_{22} - L_{02}$; $L_{24} = A_1\alpha_2 + A_2\alpha_1$. $FCR_p^c = L_{32}$. Other parameters are given in table V. *B* Three-compartment VLDL-B model [21]. Compartments 2 and 1 represent two labeled moieties of LDL-B in plasma. In the analysis it is assumed that the initial labeling of the two moieties is in proportion to their calculated masses. Although the model parameters can be solved for by fitting the data to a sum of three exponentials, and assigning them to the appropriate subsystems (one exponential to compartment 1 and two to the others), it is difficult to do this with confidence without the urine excretion data, especially the early data. It is easiest to model this on the computer and include the urine data in the simultaneous fitting of all the data.

Table V. Parameter values for LDL-B subsystem for several models

Parameter	For all models	Model A (2 compartments)	Model B (3 compartments)	Integral method
C_p (measured), mg/ml	0.76			
V_p (estimated), ml	2,870			
M_p, mg	2,180			
M_T, mg		3,159	2,698	3,159
FCR_p^*, d^{-1}	0.42			
FCR_p^c, d^{-1}		0.42	0.45	0.42
PR_p, mg/d		916	983	916
U_1, mg/d		916	448	
U_5, mg/d			535	
U_T, mg/d		916	983	916

Tracer (exogenously labeled LDL-B with iodine) administered to plasma as unit bolus. Plasma curve described by $f_p(t) = 0.706\, e^{-0.739t} + 0.294\, e^{-0.206t}$, where t is in days. $\tilde{f} = 2.38$d $= 1/0.420$; $\dot{f}(0) = 0.582/d$, and $\tilde{tf}_p = 8.22\, d^2$.

other moiety requires two compartments (compartments 2 and 4). The two moieties could be due either to two labeling sites on LDL particles or to two populations of LDL particles. The parameter values for this model are given in figure 11B. Evidence for more complex LDL-B kinetics has previously been presented by *Hurley and Scott* [29].

The FCR value for the two-moiety model depends on the assumptions made as to the nature of the two labeled moieties. Parameter values for the case of two LDL moieties and initial label on each moiety proportional to its mass are given in table V. As can be seen, these values differ from those for model A. The integral-equation method for the calculation of FCR and U for model B is incorrect, since the initial conditions do not parallel the tracee input pathways. So far, there has been no direct experimental evidence for the existence of two moieties of LDL-B.

An assumption that the two labeled moieties on LDL are two labeling sites on the same molecule leads to quite different steady-state solutions.

LDL-TG

LDL-TG data for several patients are shown in figure 12, and a model for this has recently been proposed by *Malmendier and Berman* [35] and is shown in figure 13. In this model, TG first enters with newly produced LDL particles,

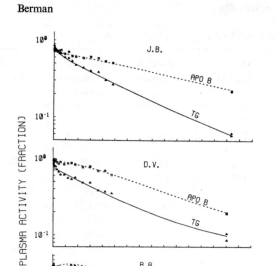

Fig. 12. Plasma LDL-TG (and LDL-B) curves after the injection of a bolus of endogenously labeled LDL-TG obtained from a donor. The model proposed for this [35] is shown in figure 13. A typical VLDL-TG curve can be described by $f_p(t) = 0.925\, e^{-7.2t} + 0.0636\, e^{-1.77t} + 0.01187\, e^{-0.1047t}$, where t is in days.

compartment 1. Most of the TG, however, is rapidly lost from this TG-rich LDL particle by delipidation ($L_{0,1}$) and the rest ($L_{2,1}$) remains with the LDL particle and undergoes the same kinetics as LDL-B, including degradation, as represented by the conventional two-compartment LDL-B model (compartments 2 and 3). The implication of this model is that there are two populations of LDL particles: one TG-rich and the other TG-poor. The TG-rich population accounts for about 10% of the number of particles and about 50% of the total LDL-TG. The TG-rich LDL particles do not seem to exchange with an extraplasma compartment (C3) while the TG-poor LDL particles do. The FCR and steady-state values for LDL-TG models are given in table VI.

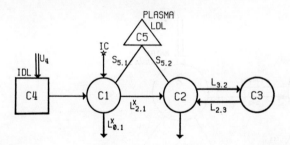

Fig. 13. LDL-TG model [35]. Plasma LDL-TG is represented by two populations: compartments 1 and 2. Compartment 1 is newly produced, TG-rich LDL particles. Most of the TG in this state is rapidly hydrolyzed. The remainder (~5–20%) becomes a more stable component of the LDL particle. FCR_p for particles entering compartment 1 is $L_{11} \cdot L_{02}/(L_{02}+L_{21})$. For TG $L_{21} \approx L_{02}$ and $FCR_p^{TG} = 0.5\, L_{11}$ (for apoB, $FCR_p^B \approx L_{02}$).

Table VI. Parameter values for LDL-TG subsystem for two models

Parameter	For all models	Three-compartment model [35]	Integral method
C_p (measured), mg/ml	0.30		
V_p (estimated), ml	2,870		
M_p, mg	861		
M_T, mg		1,871	12,500
FCR_p^*, d^{-1}	3.60		
FCR_p^c, d^{-1}		3.60	3.60
PR_p, mg/d		3,100	3,100
U_1, mg/d		3,100	3,100
U_T, mg/d		3,100	3,100

Tracer (endogenously labeled LDL-TG) administered to plasma as unit bolus [35]. Plasma curve described by $f_p(t) = 0.925\, e^{-7.2t} + 0.0636\, e^{-1.77t} + 0.01187\, e^{-0.1047t}$, where t is in days. $\dot{f}_p(0) = 6.77/d^{-1}$; $\tilde{f}_p = 0.278\, d = 1/3.60\, d$; $\widehat{f_p t} = 1.121\, d^2$, and $\widehat{f_p t}/\tilde{f}_p = 4.032\, d$.

HDL-A

Exogenously labeled (radioiodine) HDL-A, when injected into plasma as a bolus, shows a biexponential plasma activity curve [18]. Careful analysis both of plasma and urinary radioiodine excretion data using a two-compartment HDL-A model and an iodide pool (compartment 3; fig. 14) leads to the conclusion that degradation is required both from plasma and extraplasma HDL-A compartments. The need for two sites of degradation is further em-

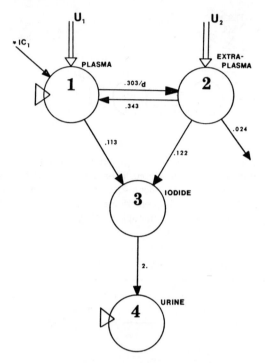

Fig. 14. Two-compartment HDL-A model [18]. Degradation of HDL-A is required from both the plasma (compartment 1) and extraplasma (compartment 2). Site of entry of new tracee material is not known. $FCR_p = L_{31} + L_{21} \cdot (L_{32} + L_{02})/L_{22}$. The model applies to the kinetics of both HDL-AI and HDL-AII, although the steady-state tracee levels and their synthesis rates may vary independently.

phasized by perturbation studies involving changes in diets and the administration of nicotinic acid [18]. These perturbations produce shifts in the degradation rates at the two sites in a complementary way, increasing one while decreasing the other, and suggests that degradation at the two sites is correlated through a common factor.

In the studies of *Blum et al.* [18], there seems to be no significant difference in the kinetics of A-I vs. A-II. The data suggest, however, that there are independent controls in the rate of synthesis of the two apoproteins.

It is not possible to resolve from present studies where the sites of entry are for newly synthesized apoA. If the site of entry of new material is compartment 1 (i.e. $U_2 = 0$), then the tracer and tracee have identical input pathways and the integral-equation method of analysis for FCR_p, PR_p, and U_p

Table VII. Parameter values for HDL-A subsystem

Parameter	For all models	Two-compartment model [18]	Integral method
C_p (measured), mg/ml	1.03		
V_p, ml	2,722		
M_p, mg	2,793		
M_T, mg		$4,522 \leq M_T \leq 6,181$	4,000
FCR_p^*, d^{-1}	0.204		
FCR_p^c, d^{-1}		0.204	0.204
PR_p, mg/d		568	568
U_p, mg/d		$568 \geq U_p \geq 0$	
U_2, mg/d		$0 \leq U_2 \leq 811$	
U_T, mg/d		$568 \leq U_T \leq 811$	

Tracer administered into plasma (compartment 1) as unit bolus of exogenously labeled HDL-A [18]. Plasma tracer curve described by $f_p(t) = 0.444\,e^{-0.777t} + 0.556\,e^{-0.1281t}$, where t is in days. $\tilde{f} = 4.91\,d$; $\dot{f}_p(0) = 0.416/d$; $\tilde{tf}_p = 34.6\,d^2$, and $\tilde{tf}_p/\tilde{f}_p = 7.05\,d$.

(equation 17) yield the same values as the compartmental model, but not M_T, the total mass of the system. If the site of entry is compartment 2, then only FCR_p and PR_p for the tracee yield the same values for the two methods of analysis, as shown in table VII.

Total System Kinetics

ApoB System

A model for total plasma apoB system is shown in figure 15 [13, 44]. It includes a synthesis pathway and VLDL-B, IDL-B and LDL-B subsystems. An amino acid subsystem is also implied, as shown by *Phair et al.* [45]. To establish the interconnections between the subsystems, experiments are required in which various products are followed after the administration of different labeled substances. This permits the establishment of necessary and sufficiency conditions for precursor-product relationships. Of particular attention in this model is the suggestion that not all LDL-B necessarily derives from VLDL-B and that not all the VLDL-B is converted to LDL-B. A detailed discussion of this model is presented by *Phair et al.* [44] and by *Berman et al.* [13].

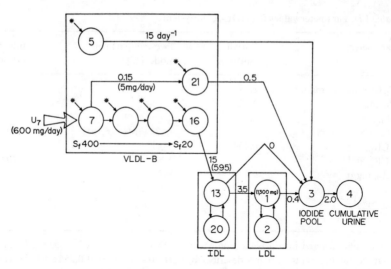

Fig. 15. Total apoB kinetics model [13,44]. ApoB is sequentially passed from VLDL to IDL to LDL. Not all apoB is synthesized via VLDL, nor does all VLDL-B pass to IDL or LDL.

ApoC System

ApoC exchanges between the various lipoprotein particles and consequently the kinetics reflect in part the kinetics of the particles and in part the exchange kinetics. Because of the multiple exchanges, apoC has to be considered as a single system that includes VLDL-C, HDL-C, IDL-C and LDL-C. This is shown in part by the model in figure 16, where the apoC states are structured the same way as apoB and TG [13]. The sites of entry of newly synthesized apoC is still not fully resolved, although the data tend to favor first appearance of apoC with HDL. The model also does not account for the individual apoC proteins C-I, C-II and C-III.

To determine the FCR and production rates for apoC, the integral method may be used if the tracer can be introduced in a manner that parallels the tracee input. This is probably not the case when apoC is exogenously labeled, especially in the case of VLDL-C where the labeling is proportional to the masses in the different plasma states. In analyzing the apoC kinetics, the imposition of constraints from the apoB and TG subsystems is most helpful.

The model suggests that only newly synthesized VLDL (and IDL) particles pick up apoC from plasma HDL and that apoC is recycled back to HDL as the VLDL particles are delipidated. Some of the exogenously labeled

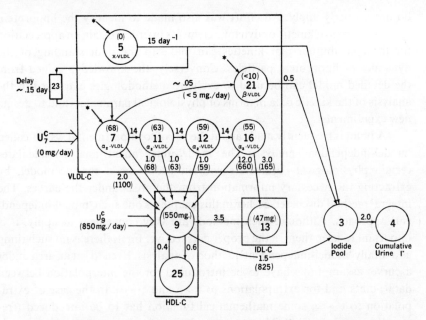

Fig. 16. ApoC kinetic model [13, 44]. Composite model for apoC kinetics based on joint apoB and apoC kinetic studies following a bolus injection of exogenously labeled (iodine) VLDL-B and VLDL-C. From the given data the site of entry of newly synthesized apoC could not be resolved. The model suggests that the exchange of apoC between VLDL and HDL is cyclic in that newly synthesized VLDL particles tend to pick up apoC from HDL and release it back to HDL during delipidation. Some apoC is also on LDL although the extent of this and the exchange process between LDL and HDL has not been fully elucidated. Compartment 5 is a rapidly disappearing component of VLDL-C which reenters the VLDL density range after a delay of about 8 h.

VLDL-C (compartment 5) is rapidly lost from plasma and reappears as VLDL-C after about a 4-hour delay. The significance of this delay is not clear. More detailed discussion of the apoC model appears elsewhere [13, 44].

Discussion

Some basic concepts underlying the analysis of kinetic data and their interpretation in terms of physiological parameters have been described. It was shown how some simplified methods of analysis could be employed for special cases and how these methods fail when the underlying assumptions do not properly apply. An effort was also made to identify the lipoprotein system in terms of kinetic or dynamic elements in order to gain an appreciation for the contribution that kinetics can make to the understanding of the dynamics of lipoprotein particles. Conversely, the knowledge gained from the detailed model can be used to evaluate methodologies employed in the analysis of the kinetic data in terms of physiological parameters and to design new experiments.

A point is frequently made about the use of 'compartmental' vs. so-called 'model-independent' analysis. The 'model-independent' analysis calculates certain physiological parameters without proposing a specific model, by extracting the necessary information from the areas under the curves. The integral method discussed earlier in this chapter is one such 'model-independent' approach. Although such methods are very convenient to apply, it is important to keep them in the proper perspective. First, there is no such thing as a totally 'model-independent' method of analysis. Even to get an area under a curve, assumptions have to be introduced for the interpolation between data points and for extrapolations to $t \to 0$ and $t \to \infty$. In the case of extrapolation to $t \to \infty$, some mathematical function has to be introduced (frequently an exponential). This arbitrary choice of an extrapolation function is a hypothesis and the function chosen implies a model. When an exponential is chosen, the last few data points are used to determine its decay rate. Such a procedure tends to underestimate areas under the curves. Worse than this, the extent of the underestimation is not known, hence the calculated value is more properly an estimate of a lower limit of the area, rather than the area. In addition, the use of an area ignores fine details of a curve, thus discarding information. The method also cannot make use of isolated data that do not constitute 'areas'. The use of a specific model or compartmental analysis, therefore, has definite advantages when the system is not known. A detailed model extracts all the information in the data. If data contain redundant information, it serves to validate further the model. Once a model is tested for consistency, it can be used for the evolution of the adequacy of simpler methods of analysis. Finally, it should be realized that compartmental models must yield the same result as simpler methods of analysis when the use of the latter is valid. In short, in order to learn as much as possible about a system, validate a model to the greatest extent and identify the areas that require further experimentation, detailed modeling is essential. Simpler methods of analysis may be taken advantage of for special purposes and for routine calculation once their use is justified after the testing and integration of all available information.

Deconvolution is a powerful technique in that it permits the calculation of an input function to a subsystem, given the responses of the subsystem to a unit impulse and to the unknown input. It has been used, for example, by *Sigurdsson et al.* [53–55] to calculate the LDL-B input function from a VLDL-B precursor and thus to determine which fraction of VLDL-B goes to LDL-B. It is critical in using this technique that the assumptions underlying deconvolution be fully satisfied. In the case of VLDL-LDL interconversion, a question may be raised as to the validity of the method. One of the assumptions is that the input function follow the exact paths as the unit tracer input. In the case of LDL-B, the tracer input is achieved by exogenous labeling. As pointed out earlier, there is suggestive evidence that the LDL-B label is *not* homogeneous. Whether this is an artifact of labeling or whether it is due to two moieties of LDL is not known. In either case, it is not likely that LDL-B derived endogenously from a VLDL-B precursor is labeled in a similar way. Furthermore, VLDL-B, when used as a precursor, is also not homogeneously labeled, which makes the deconvolution results even more questionable. It is critical that the details of the VLDL-B and LDL-B subsystems be fully resolved before the results of deconvolution techniques can be accepted with confidence.

Independently of the above considerations, deconvolution is a very sensitive numerical procedure and can readily generate erroneous results. It is particularly sensitive to noisy data (derivatives) and the estimate of the initial slope of a curve. These are frequently difficult to evaluate because of sparse initial data and the presence of contaminants and artifacts as a result of experimental limitations.

I would like to reemphasize the use of exogenous labeling as a tracer for the tracee. When more than a single population is present, exogenous labeling is at best, given similar efficiencies of labeling, proportional to the masses of the populations, and not the rates of inputs of tracee into these populations. Under these conditions, the tracer kinetics can be properly related to the tracee only if the details of the subsystem and the labeling techniques are known. In connection with the latter, it should be realized that efficiency of labeling (tracer atoms/tracee molecule) and the form of label (e. g. ^{14}C vs. ^{3}H) and the way it is introduced (exogenous vs. endogenous) can further complicate the interpretation of tracer results in terms of the tracee.

Two considerations have been examined and stressed here in the analysis of VLDL and LDL data: multiple populations and a delipidation cascade. It was shown that these can have significant effects on the interpretation of the data, especially the equivalences between a tracer and tracee. It points up the fact that VLDL-B and VLDL-TG should have the same FCR if delipidation is a single-step process as far as plasma is concerned, but different FCRs if multiple delipidation steps are involved. It is important in future work that attention be given to these considerations so that changes in metabolic patterns be correctly interpreted.

FCR has been used here as a measure of metabolism because of its common usage in the literature. It should be realized that FCR is commonly defined with respect to an initial distribution space, usually defined by plasma or a plasma-containing compartment. The initial space of distribution can be calculated from the initial dilution of the tracer, however, the data are frequently not good enough to define such a space. The presence of experimental artifacts further complicates this determination. Unfortunately, the value of FCR depends directly on the initial distribution space. For this reason it may be advisable to use a metabolic clearance rate which is independent of the initial distribution volume. Or else, when in doubt, an estimated plasma volume may be imposed as a constraint for the initial space of distribution so that it may serve as a standard reference for intercomparisons.

Although I have tried to develop a 'total' kinetic scheme for the various moieties in the different lipoprotein particles, with the point of view that the particles serve as a common carriers, I have succeeded only partially. The reason for this is the fact that in developing the models for each subsystem, multiple populations of particles have emerged and there are not enough data yet to indicate how these populations are generated within each lipoprotein class or from precursors. For example, LDL seems to require two populations of labeled particles to explain its apoB kinetics following exogenous labeling. Are the two populations physiological? If so, do these come from VLDL-B? These questions need be answered before a fully integrated model can be presented. The same applies to the triglyceride and apoC moieties. Cholesterol and phospholipids have not been dealt with here, but they too have to fit the overall kinetic patterns of the lipoprotein particles.

References

1 Alaupovic, P.; Lee, D.M., and McConathy, W.J.: Studies on the composition and structure of plasma lipoproteins. Distribution of lipoprotein families in major density classes of normal human plasma lipoprotein. Biochim. biophys. Acta *260:* 689–707 (1972).
2 Baker, N.: The use of computers to study rates of lipid metabolism. J. Lipid Res. *10:* 1–24 (1969).

3 Barter, P.J. and Nestel, P.J.: Precursor-product relationship between pools of very low density lipoprotein. Biochim. biophys. Acta 260: 212–221 (1972).
4 Barter, P.J. and Nestel, P.J.: Precursor-product relationship between pools of very low density lipoprotein triglyceride. J. clin. Invest. 51: 174–180 (1972).
5 Bergner, P.E.E.: Exchangeable mass determination without assumption of isotopic equilibrium. Science 150: 1048–1050 (1965).
6 Berman, M.: A deconvolution scheme. Mathl Biosics 40: 319–323 (1978).
7 Berman, M.: A postulate to aid in model building. J. theor. Biol. 4: 229–236 (1963).
8 Berman, M.: Compartmental analysis in kinetics; in Stacy and Waxman, Computers in biomedical research, vol. 2, chap. 7 (Academic Press, New York 1965).
9 Berman, M.: Compartmental modeling; in Laughlin and Webster, Advances in medical physics, pp. 199–296 (Second International Conference on Medical Physics, Inc., Boston 1971).
10 Berman, M.: Iodine kinetics; in Rall and Kopin, Methods of investigative and diagnostic endocrinology, pp. 172–203 (North-Holland, Amsterdam 1972).
11 Berman, M.: Kinetic modeling in physiology. FEBS Lett. 2: suppl., pp. 556–557 (1969).
12 Berman, M.: The formulation and testing of models. Ann. N.Y. Acad. Sci. 108: 182–194 (1963).
13 Berman, M.; Hall, M. III; Levy, R.I.; Eisenberg, S.; Bilheimer, D.W.; Phair, R.D., and Goebel, R.H.: Metabolism of apoB and apoC lipoproteins in man: kinetic studies in normal and hyperlipoproteinemic subjects. J. Lipid Res. 19: 38–56 (1978).
14 Berman, M. and Schoenfeld, R.: Invariants in experimental data on linear kinetics and the formulation of models. J. appl. Phys. 27: 1361–1370 (1956).
15 Berman, M. and Weiss, M.F.: SAAM manual. US DHEW publ. No. (NIH), 78–180 (1978).
16 Berman, M.; Weiss, M.F., and Shahn, E.: Some formal approaches to the analysis of kinetic data in terms of linear compartmental systems. Biophys. J. 2: 289 (1962).
17 Bilheimer, D.W.; Goldstein, J.L.; Grundy, S.M., and Brown, M.S.: Reduction in cholesterol and low density lipoprotein synthesis after portocaval shunt surgery in a patient with homozygous familial hypercholesterolemia. J. clin. Invest. 56: 1420–1430 (1975).
18 Blum, C.G.; Levy, R.I.; Eisenberg, S.; Hall, M. III; Goebel, R.H., and Berman, M.: High density lipoprotein in man. J. clin. Invest. 60: 795–807 (1977).
19 Eaton, R.P. and Kipnis, D.M.: Incorporation of Se^{75} selenomethionine into a protein component of plasma very-low-density lipoprotein in man. Diabetes 21: 774–753 (1972).
20 Farquhar, J.W.; Gross, R.C.; Wagner, R.M., and Reaven, G.M.: Validation of an incompletely coupled two-compartment nonrecycling catenary model for turnover of liver and plasma triglyceride in man. J. Lipid Res. 6: 119–134 (1965).
21 Goebel, R.; Garnick, M., and Berman, M.: A new model for low density apoprotein kinetics: evidence for two labeled moieties. Circulation 54: suppl. II, pp. II–4 (1976).
22 Groth, T.: Biomedical modeling; in Shires and Wolf, Medinfo 77, pp. 775–784 (North-Holland, Amsterdam 1977).
23 Grundy, S.M. and Mok, H.M.: Chylomicron clearance in normal and hyperlipidemic man. Metabolism. 25: 1228–1239 (1976).
24 Gurpide, E.: Tracer methods in hormone research; in Gross, Labhart, Lipsett,

Mann, Samuels and Zander, Monographs on Endocrinology (Springer, New York 1975).

25 Hart, H.E.: Analysis of tracer experiments. VII. General multi-compartment systems imbedded in non-homogeneous inaccessible media. Bull. math. Biophys. *28:* 261 (1966).

26 Hart, H.E.: Analysis of tracer experiments. VIII. Integro-differential equation treatment of partly accessible, partly injectable multicompartment systems. Bull. math. Biophys. *29:* 319–333 (1967).

27 Havel, R.J. and Kane, J.P.: Quantification of triglyceride transport in blood plasma: a critical analysis. Fed. Proc. Fed. Am. Socs exp. Biol. *34:* 2250–2257 (1975).

28 Hearon, J.Z.: Theorems on linear systems. N.Y. Acad. Sci. *108:* 36–38 (1963).

29 Hurley, P.J. and Scott, P.J.: Plasma turnover of S_f 0–9 low-density lipoprotein in normal men and women. Atherosclerosis *11:* 51–76 (1970).

30 Jacquez, J.A.: Compartmental analysis in biology and medicine – kinetics of tracer-labeled materials, pp. 1–79, 121–142 (Elsevier, Amsterdam 1972).

31 Kellershohn, C.: New formulation of earlier results in the kinetic theory of tracer application to various problems in metabolism; in Laughlin and Webster, Advances in medical physics, pp. 141–163 (Second International Conference on Medical Physics, Inc., Boston 1971).

32 Kekki, M. and Nikkila, E.A.: Turnover of plasma total and very low density lipoprotein triglyceride in man. Scand. J. clin. Lab. Invest. *35:* 171–179 (1975).

33 Knott, G.: MLAB: an online modeling laboratory (Division of Computer Research Technology, NIH, Bethesda (1975).

34 Langer, T.; Strober, W., and Levy, R.I.: The metabolism of low density lipoprotein in familial type II hyperlipoproteinemia. J. clin. Invest. *51:* 1528–1536 (1972).

35 Malmendier, C.L. and Berman, M.: Endogenously labeled low density lipoprotein triglycerides and apoprotein B kinetics. J. Lipid Res. *19:* 978–984 (1978).

36 Malmendier, C.L.; Delcroix, C., and Berman, M.: Interrelations in the oxidative metabolism of free fatty acids, glucose and glycerol in normal and hyperlipemic patients. J. clin. Invest. *54:* 461–476 (1974).

37 Matthews, C.M.E.: The theory of tracer experiments with 131-I labelled plasma proteins. Physics Med. Biol. *2:* 36–53 (1957).

38 Meier, P. and Zierler, K.L.: On the theory of the indicator-dilution method for measurement of blood flow and volume. J. appl. Physiol. *6:* 731–744 (1954).

39 Metzler, C.M.: Biostatistical technical report No. 7292–69–7292–005 (Upjohn, Kalamazoo 1969).

40 Nikkila, E.A. and Kekki, M.: Polymorphism of plasma triglyceride. Kinetics in normal human adult subjects. Acta med. scand. *190:* 49–59 (1971).

41 Nosslin, A.: Appendix in Andersen, Metabolism of human gamma-globulin, pp. 115–120 (Blackwell, Oxford 1964).

42 Osborne, J.C. and Brewer, H.B.: The plasma lipoproteins. Adv. Protein Chem. *31:* 253–337 (1977).

43 Perl, W. and Samuel, P.: Input-output analysis for total input rate and total traced mass of body cholesterol in man. Circulation Res. *25:* 191–199 (1969).

44 Phair, R.D.; Hall, M. III; Bilheimer, D.W.; Levy, R.I.; Goebel, R.H., and Berman, M.: Modeling lipoprotein metabolism in man; in 1976 Summer Computer Simulation Conference, pp. 486–489 (California Simulation Councils, Inc., 1976).

45 Phair, R.D.; Hammond, M.G.; Bowden, J.A.; Fired, M.; Fisher, W.R., and Berman, M.: Preliminary model for human lipoprotein metabolism and hyperlipoproteinemia. Fed. Proc. Fed. Am. Socs exp. Biol. *34:* 2263–2270 (1975).

46 Reaven, G.M.; Hill, D.B.; Gross, R.C., and Farquhar, J.W.: Kinetics of triglyceride turnover of very low density lipoproteins of human plasma. J. clin. Invest. *44:* 1826–1833 (1965).

47 Rescigno, A. and Segre, G.: Drug and tracer kinetics (Blaisdell, Waltham 1966).

48 Rescigno, A. and Gurpide, E.: Estimation of average times of residence, recycle and interconversion of blood-borne compounds using tracer methods. J. clin. Endocr. Metab. *36:* 263–276 (1973).

49 Shames, D.M.; Frank, A.; Steinberg, D., and Berman, M.: Transport of plasma free fatty acids and triglycerides in man: a theoretical analysis. J. clin. Invest. *49:* 2298–2314 (1970).

50 Sheppard, C.W.: Principles of the tracer method (Wiley, New York 1961).

51 Shipley, R.A. and Clark, R.E.: Tracer methods for *in vivo* kinetics, theory and applications (Academic Press, New York 1972).

52 Shore, V.G. and Shore, B.: Heterogeneity of human plasma very low density lipoproteins. Separation of species differing in protein components. Biochemistry, N.Y. *12:* 502–507 (1973).

53 Sigurdsson, G.; Nicoll, A., and Lewis, B.: Conversion of very low density lipoprotein to low density lipoprotein. A metabolic study of apoprotein B kinetics in human subjects. J. clin. Invest. *56:* 1481–1490 (1975).

54 Sigurdsson, G.; Nicoll, A., and Lewis, B.: Metabolism of very low density lipoproteins in hyperlipidaemia: studies of apolipoprotein B kinetics in man. Eur. J. clin. Invest. *6:* 167–177 (1976).

55 Sigurdsson, G.; Nicoll, A., and Lewis, B.: The metabolism of low density lipoprotein in endogenous hypertriglyceridaemia. Eur. J. clin. Invest. *6:* 151–158 (1976).

56 Skinner, S.M.; Clark, R.E.; Baker, N., and Shipley, R.A.: Complete solution of the three-compartment model in steady state after single injection of radio-active tracer. Am. J. Physiol. *196:* 238–244 (1959).

57 Sniderman, A.D.; Carew, T.E.; Chandler, J.G., and Steinberg, D.: Paradoxical increase in rate of catabolism of low density lipoproteins after hepatectomy. Science *183:* 526–528 (1974).

58 Steiner, G. and Streja, D.: Kinetics of VLDL subfractions; in Schettler, Goto, Hata and Klose, Atherosclerosis, IV, pp. 129–132 (Springer, Berlin 1977).

59 Stephenson, J.L.: Theory of transport in linear biological systems. I. Fundamental integral equation. Bull. math. Biophys. *22:* 1–17 (1960).

60 Stephenson, J.L.: Theory of transport in linear biological systems. II. Multiflux problems. Bull. math. Biophys. *22:* 113–138 (1960).

61 Stewart, G.N.: Pulmonary circulation time: the quantity of blood in the lungs and the output of the heart. Am. J. Physiol. *58:* 20 (1921).

62 Tait, J.F.: Review: The use of isotopic steriods of production rates *in vivo*. J. clin. Endocr. Metab. *23:* 1285–1297 (1963).

63 Thompson, G.R.; Spinks, T.; Ramicar, A., and Myant, N.B.: Non-steady-state studies of low-density-lipoprotein turnover in familial hypercholesterolaemia. Clin. Sci. mol. Med. *52:* 361–369 (1977).

64 Zech, L.A.; Grundy, S.M.; Steinberg, D., and Berman, M.: A kinetic model for production and metabolism of very low density lipoprotein triglycerides: evidence for a slow production pathway and results for normolipidemic subjects (submitted).
65 Zilversmit, D.B.; Entenman, C., and Fishler, M.C.: Calculation of turnover time and turnover rate from experiments involving the use of labeling agents. J. gen. Physiol. *26:* 325–331 (1943).

M. Berman, PhD, Laboratory of Theoretical Biology, DCBD, NCI,
National Institutes of Health, Bethesda, MD 20014 (USA)

Chylomicrons
Mechanism of Transfer of Lipolytic Products to Cells

Louis C. Smith and Robert O. Scow

Division of Atherosclerosis and Lipoprotein Research, Department of Medicine, Baylor College of Medicine and The Methodist Hospital, Houston, Tex., and Section on Endocrinology, Laboratory of Nutrition and Endocrinology, National Institute of Arthritis, Metabolism and Digestive Diseases, Bethesda, Md.

Introduction

Triacylglycerol in the diet is usually the major source of the daily flux of fatty acid in the circulation. Fatty acids and monoacylglycerol are released from dietary triacylglycerol by enzymic hydrolysis in the stomach and intestines. They are absorbed by the intestinal epithelium, esterified to triacylglycerol and combined with cholesterol, acylcholesterol, phospholipids and specific apolipoproteins into triacylglycerol-rich lipoproteins, chylomicrons and very low density lipoproteins (VLDL). Through a complex series of processes which are poorly understood, these particles are released into the circulation for transport of triacylglycerol to individual tissues [1–6].

The lipoproteins that transport triacylglycerol, when examined by scanning electron microscopy, appear as spheres with smooth surfaces and diameters between 0.03 and 0.6 μm [8]. One class of particles, chylomicrons, has a density less than 0.95 g cm^{-3} and a mass between 0.4 and 30×10^9 daltons [7], while the other major class, VLDL, has a density between 0.95 and 1.006 g cm^{-3} and a narrow mass range, $5–10 \times 10^6$ daltons. Both density classes are formed by the intestine. Although they may differ in as yet undetermined ways, they are both designated for convenience as 'chylomicrons' in this article. All evidence is consistent with a structure in which a triacylglycerol core is surrounded by phospholipid, mostly as phosphatidylcholine, cholesterol and apolipoproteins that are spread as a monomolecular layer on the surface of the lipoprotein particle [8,9]. Irrespective of the size, the ratio of triacylglycerol to phospholipids shows a highly significant correlation with the ratio of surface to volume of chylomicrons [9]. The chylomicron core appears circular, homogeneous, and smooth at the periphery in thin sections of specimens fixed with OsO_4 [10]. Triacylglycerol and acylcholesterol are miscible [11] and constitute the core of the lipoprotein, although a small amount of triacylglycerol may also be present in the surface film. Because the molecules contain a hydrophilic head group, fatty acid, mono- and diacylglycerol presumably reside also at the lipoprotein surface. Similarly, the polar 3β-hydroxyl group of cholesterol and the interactions of cholesterol with phospholipids confine the sterol primarily in the surface film. However, since cholesterol is sparingly soluble in core lipids [12, 13], it may also be found in the core.

Triacylglycerol in chylomicrons is removed rapidly from the circulation, with a half-life less than 5 min in man [14], as the result of lipoprotein lipase catalysis in capillary endothelium [3, 15]. The level of activity of this enzyme in individual tissues, such as adipose tissue, heart, muscle and mammary gland, varies with nutritional and physiological states that affect triacylglycerol uptake, such as fasting, exercise, pregnancy and lactation [3, 4]. Chylomicron triacylglycerol is not utilized directly by liver. However, when triacylglycerol is hydrolyzed in peripheral tissues by lipoprotein lipase, some of the fatty acids formed are released to the blood and taken up by the liver.

Physical transfer of fatty acid moieties from the triacylglycerol-rich lipoproteins to individual cells requires at least three processes: an enzymatic cleavage of an ester bond, physical transfer of the monomolecular lipolytic products and conversion to another chemical form inside the cell. The kinetics, mechanisms and control of these processes have received little attention, because the dynamic character of lipoprotein and membrane structure has been described only recently [16, 17]. It has become apparent that lipoproteins and membranes are not simply static systems of stoichiometric complexes which determine transport functions by distance, orientation and organization of the components. These lipid-protein surface have the dynamic properties of a two-dimensional fluid, in which the functions are determined primarily by rapid noncovalent interactions between individual lipid and protein components. These molecules adsorb and desorb from the surfaces, diffuse in both a rotational and translational sense in the surfaces, and move from one side of the membrane to the other, in the case of bilayer membranes. In the context of this dynamic structure of lipoproteins and membranes, we proposed earlier [18, 19] a mechanism of transfer of lipolytic products from the chylomicrons involving fusion of the surface film of the chylomicron with the external leaflet of the endothelial cell and subsequent transfer of monoacylglycerol and fatty acid by lateral movement across the endothelial cell through an interfacial continuum of plasma and intracellular membranes. The focus of this article is to develop the rationale and experimental support for this and other postulated mechanisms and to speculate, on the basis of presently available information, about the feasibility of each transfer mechanism. Other aspects of lipoprotein lipase and triacylglycerol metabolism can be found in other chapters of this book and in recent reviews [1,20,21].

Role of Lipoprotein Lipase

Phase Transfer of Fatty Acyl Groups

Triacylglycerol is poorly soluble in phosphatidylcholine [22] and might be present as a component of the chylomicron surface film. As such, it would be in equilibrium with the surrounding aqueous phase as well as with the apolar core. The limited solubility of triacylglycerol in the interface and in the aqueous phase accounts for its pronounced tendency to form a separate phase, the chylomicron core. The extremely low solubility of the long-chain triacylglycerol in aqueous solution makes it unlikely that there would be any detectable uptake of intact individual triacylglycerol molecules from the aqueous phase at the endothelial cell surface. Thus, the equilibrium of the triacylglycerol between the lipoprotein and the aqueous phase is not important, in either a thermodynamic or kinetic sense, to the overall transfer process [23, 24].

Physical movement of fatty acyl groups out of the apolar core of chylomicrons into tissue requires the action of lipoprotein lipase to transform the hydrophobic triacylglycerol into more polar acyl lipids. The 1(3)-ester of triacylglycerol is cleaved by lipoprotein lipase, releasing fatty acid. The other initial product, *sn*-2,3(1,2)-diacylglycerol, is then hydrolyzed to *sn*-2-monoacylglycerol and fatty acid. The equilibrium distribution of fatty acyl groups between the core phase, the interfacial phase and the aqueous phase is drastically altered when their chemical form, and consquently their physical properties, is changed. The lipolytic products, due to their strong amphipathic character, will locate primarily in the interface [25]. Thus, the role of lipoprotein lipase is to alter the chemical form of fatty acyl groups, while transfer from the site of hydrolysis occurs primarily by simple physical processes described later in this article. While most of the fatty acyl groups transferred to the tissue originate from triacylglycerol, some fatty acids also come from hydrolysis of phospholipid in the surface film [26, 27].

Interaction with Apolipoprotein C-II

The activation of lipoprotein lipase by apolipoprotein C-II (apoC-II), a protein component of the surface film of the chylomicron and VLDL, has been known for some years [28, 29], although the precise mechanism by which this activation occurs has not yet been established. Lipoprotein lipase does hydrolyze triacylglycerol in the absence of apoC-II, thus, excluding a true coenzyme function for the apolipoprotein. As a working hypothesis, we assume there is a specific protein-protein interaction between apoC-II in the lipoprotein substrate and lipoprotein lipase at the endothelial cell surface. Initial support for this postulate has been obtained with monomolecular surface films of apoC-II [30]. ApoC-II forms a stable surface film, in the absence of lipid, with a conformation that is biologically active, as shown by the ability of the isolated apoliprotein surface film to activate lipoprotein lipase. When apoC-II and lipoprotein lipase are allowed to interact in the absence of lipid, a stable apolipoprotein: enzyme complex forms at the interface. This surface film complex can be isolated by physical transfer from one subphase to another. By other techniques, this association of lipoprotein lipase and apoC-II has been shown to be extremely strong. The calculated equilibrium dissociation constant for the enzyme: apoC-II complex is $< 10^{-8}$ M [31] from one laboratory and 3×10^{-13} M from another [32]. The rate of triacylglycerol hydrolysis increases as a function of apoC-II concentration and is maximal at a molar ratio of enzyme to apoC-II of 1:1 [31, 32]. Formation of this surface film complex is prevented by sodium chloride [30]. Inhibition of lipoprotein lipase activation is anion specific [33]; we suggest that the inability to activate results from the failure to form this enzyme: apolipoprotein complex.

Kinnunen et al. [34] have identified three different regions of the sequence of apoC-II that are involved in apoC-II activation of lipoprotein lipase. Cyanogen bromide (CNBr) fragments of apoC-II containing residues 1-9 and 10-59 do not activate. The carboxyl-terminal CNBr fragment corresponding to residues 60-78 increases the hydrolysis of trioleoylglycerol in a gum-arabic-stabilized emulsion fourfold, compared to a ninefold increase by apoC-II at the same concentration. ApoC-II 60-78, a synthetic peptide prepared by solid-phase techniques, stimulates lipolysis to about the same extent as does the corresponding CNBr fragment of native apoC-II. Addition of five residues produces a synthetic peptide, apoC-II 55-78, which gives an activation essentially identical to that produced by apoC-II. Synthetic peptides, apoC-II 50-78 and apoC-II 43-78, are also equally effective as activators. By contrast, apoC-II 66–78 does not activate; removal of the three carboxyl-terminal residues, GLY-GLU-GLU, from CNBr fragment 60–78 also abolishes activation. These studies suggest that maximal activation of lipoprotein lipase by apoC-II requires a minimal sequence of 24 amino acids contained within residues 55–78. We speculate that these two carboxyl-terminal glutamic acid residues of apoC-II interact electrostatically with a positively charged region of lipoprotein lipase and that this ionic interaction is specifically disrupted by high concentrations of anions. We propose that residues 55–67 interact specifically with lipoprotein lipase to change enzyme structure and thereby enhance enzymatic catalysis.

The primary function of the amino-terminal 49 residues appears to be lipid binding. ApoC-II, like other plasma apolipoproteins, forms a stable complex with phospholipids [35]. The fluorescence emission maximum of tryptophan shifts from 349 to 330 nm and the helicity of apoC-II increases about threefold as the result of binding to dimyristoylphosphatidylcholine. The synthetic peptide apoC-II 55–78 has a disordered structure by circular dichroism [36]; neither apoC-II 55–78 nor apoC-II 50–78 can form a complex with phosphatidylcholine sufficiently stable for isolation. By contrast, addition of seven more residues produces apoC-II 43–78 that forms a stable complex with phosphatidylcholine, as indicated by the increase in α-helical content and by density gradient isolation of the complex.

A possible interaction of lipoprotein lipase and apoC-II in the chylomicron surface is depicted in figure 1. We propose that the lipid-binding regions of apoC-II, residues 1–50, are in the plane of the surface and retain apoC-II in the lipoprotein substrate. There appears to be a similar interaction between the lipoprotein surface lipids and lipoprotein lipase. Under conditions in which high concentrations of NaCl prevent activation of lipoprotein lipase by apoC-II, the enzyme remains bound to lipoprotein throughout ultracentrifugal isolation [33]. In addition, changes in tryptophan fluorescence show that lipoprotein lipase binds to dimyristoylphosphatidylcholine vesicles in the absence of apoC-II [37]. Taken together, these findings imply discrete lipid- binding regions in lipoprotein lipase. Although the exact two-dimensional arrangement of proteins at the surface of lipoproteins is not known, we assume that the lipid-binding regions of lipoprotein lipase have the same conformation at the surface as do the lipid-binding regions of apoproteins [38]. Presumably, amphipathic regions of both proteins are oriented in such a way that the ionized groups of the polar amino acid residues are in the aqueous phase and the hydrophobic portion of the nonpolar amino acid residues face the interior of the lipoprotein surface. In such a configuration,

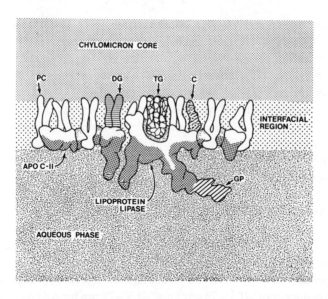

Fig. 1. Model of apoC-II interaction with lipoprotein lipase in the chylomicron surface. The hydrophilic sides of the proteins are indicated by the heavy stippled pattern. PC = Phosphatidylcholine; DG = diacylglycerol; TG = triacylglycerol; C = cholesterol; GP = glycoprotein portion of lipoprotein lipase.

activation of the lipoprotein lipase by apoC-II could by lateral interaction of residues 55–67 of apoC-II with a structural site of the enzyme near or at the active site, thus altering the catalytic efficiency of the enzyme. Since apoC-II 50–78 does not form a stable complex with phospholipid, the carboxyl-terminal sequence is depicted in the aqueous phase where there may be a relatively long-range ionic interaction between the terminal LYS-GLY-GLU-GLU and an oppositely charged region of lipoprotein lipase. Several apoC-II fragments activate lipoprotein lipase even though they do not bind lipid. This finding indicates that activation of lipoprotein lipase by apoC-II cannot result simply from alterations in the substrate structure by the apolipoprotein that permit a faster rate of penetration of the surface film by the enzyme.

Lipoprotein Lipase Action on Chylomicrons at the Endothelial Cell Surface

Capillaries in adipose and mammary tissue, as well as in heart and skeletal muscles, consist of a single layer of endothelial cells surrounded by a con-

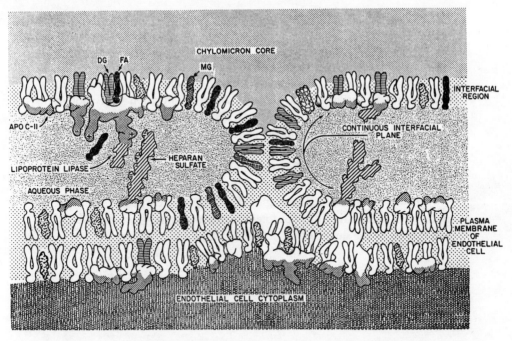

Fig. 2. Model of possible mechanisms of transfer of lipolytic products from the chylomicron to the endothelial cell. MG = Monoacylglycerol; FA = fatty acid. The designations of other molecules are the same as in figure 1.

tinuous basement membrane and pericytes [18, 19]. Since there is no morphological evidence that chylomicrons cross the endothelium, hydrolysis of triacylglycerol would have to occur at the luminal surface of the endothelium.

Lipoprotein lipase is one of several enzymes from individual tissues [3, 4, 39, 40] easily released into the plasma by injection of heparin. The enzyme apparently is a peripheral protein, perhaps loosely bound to the endothelial cell surface by glycosaminoglycans, similar to heparin [3]. *Olivecrona et al.* [41] have speculated that the lipase is attached by electrostatic interactions to heparan sulfate polymers that extend 20–50 nm from the endothelial cell surface (fig. 2). The distance between the chylomicron surface and the plasma membrane of the luminal surface of the endothelial cell could vary, depending on the number of repeating units of the glycosaminoglycans.

The sequence of events that precedes hydrolysis of chylomicron triacylglycerol, and precisely which components of the lipoprotein interact initially with endothelial cell surface, is not known. By analogy with the *in vitro* interaction of apoC-II and lipoprotein lipase, there may be a protein-protein interaction between the apolipoprotein and the enzyme to form a membrane-bound chylomicron:lipoprotein lipase complex. Lipid binding by lipoprotein lipase may also occur without interaction with apoC-II; after penetration of the lipoprotein surface by the enzyme, lateral association with apoC-II in the chylomicron surface could follow. It is also possible that the initial interaction of the chylomicron and the endothelial cell surface does not involve lipoprotein lipase. For example, apoB and apoE bind to heparin *in vitro* [42, 43]. An electrostatic interaction of the chylomicron with glycosaminoglycan chains could attach the lipoprotein to the luminal surface of the endothelial cell for subsequent interaction of apoC-II and lipoprotein lipase. As illustrated in figure 2, orientation of lipoprotein lipase and apoC-II in the interface is assumed to be the same as that depicted in figure 1.

Transfer of Fatty Acid and Monoacylglycerol from the Chylomicron Surface

Mechanisms of Lipid Transfer

As fatty acid and monoacylglycerol are produced by lipoprotein lipase, physical transfer of these lipolytic products from the site of enzyme action at the chylomicron surface can occur by several different mechanisms. Most of the experimental evidence for these mechansisms has been obtained with lipoproteins and lipid vesicles. Our working assumption is that the mechanisms for physical transfer of components between lipoproteins and cell membranes are the same as those which occur between lipoproteins, since the structure and composition of plasma lipoproteins and cell membranes have much in common [5].

Hypotheses

Three mechanisms have been proposed for exchange of lipids between lipoproteins. In the first hypotheses, a lipid molecule desorbs from the lipoprotein surface [44] into the aqueous medium and is adsorbed by other lipoproteins or membranes thus effecting a net transfer of lipid. From the kinetic point of view, this process, termed molecular diffusion, is described by two reactions: a relatively slow first-order dissociation reaction, followed by a more rapid second-order association reaction. When an identical lipid molecule transfers from the acceptor surface to the original donor, the mass balance is restored and exchange of lipid molecules has occurred by two identical independent processes.

In the second mechanism, lipid transfer involves the formation of a collision complex of particles [45] that exists sufficiently long for lipid molecules to diffuse in the complex. The temporary bridge formed by fusion [46], the first step, would permit a redistribution of lipid molecules between lipid aggregates, the second step, without the molecules having to come in direct contact with the aqueous environment. Transfer and exchange could occur simultaneously and independently. This mechanism predicts classical second-order raction kinetics.

Specific lipid transfer proteins [47–51] are the central feature of the third mechanism of transfer. The lipid may desorb from the lipoprotein surface before interaction with the specific carrier protein, or the protein may first associate with or penetrate the surface film and there interact with the specific lipid. The complex could then desorb to diffuse in the solution to another surface where the steps would be reversed to complete the transfer. Because the desorption steps and the conformational changes in protein structure that accompany lipid binding and penetration of the lipoprotein surface may have comparable rate constants, the apparent kinetics of transfer cannot be predicted.

Experimental evidence for these mechanisms is briefly summarized, followed by a description, in terms of these proposed mechanisms, of how fatty acid and monoacylglycerol might transfer from the chylomicron surface to the endothelial cell surface.

Transfer of Fluorescent Lipids by Molecular Diffusion

The exchange of phospholipids and cholesterol between lipoproteins and various blood components has been known for many years [46, 52]. However, the mechanisms of these transfers are poorly understood. Published values for the rates of transfer of many lipids and apoproteins have questionable validity since the times required to isolate the lipoproteins for analysis are much greater than the half-times of exchange. The use of high concentrations of salt [53] for ultracentrifugal separation of reactants, and of heparin-$MnCl_2$ to precipitate lipoproteins [54], creates further ambiguity concerning interpretation of the experiments since these procedures can alter both lipoprotein structure and exchange rates.

Recently, fluorescence techniques have been successfully used to avoid some of the experimental problems encountered in earlier work, as well as to elucidate the mechanism of transfer of lipids between lipoproteins. The fluorescence properties of pyrene and lipids containing a covalent pyrenyl moiety depends on the microscopic concentration of the compound [55]. Dilute solutions of pyrene exhibit a fluorescence from the lowest excited singlet state with a maximum at 390 nm. This is termed monomer fluorescence. At higher pyrene concentrations, an excited *dimer* (excimer) is produced by the transient association of an excited singlet state molecule with a ground state pyrene molecule. As the pyrene concentration increases, excimer emission at 470 nm increases at the expense of the monomer emission, a special case of fluorescence quenching. The ratio of these fluorescence intensities is directly proportional to the pyrene concentration [56]. This property of pyrene and its derivatives can be used to monitor transfer of lipids between lipoproteins when a change in concentration occurs. This strategy has been successfully applied to transfer of pyrene [57], *rac*-1-oleyl-2-[4(3-pyrenyl)-butanoyl]-glycerol, a fluorescent diglyceride analog [58], and pyrenemethyl-23,24-*dinor*-5-cholen-22-oate-3β-ol, a cholesterol analog [59], between HDL. Transfer of 10-(3-pyrenyl)-decanoic acid [60] and 9-(3-pyrenyl)-nanonoic acid (PNA) [61] between phosphatidylcholine vesicles has also been studied.

In all instances, when HDL or phosphatidylcholine vesicles containing the fluorescent probe are mixed with unlabeled acceptor, excimer fluorescence decreases and monomer fluorescence increases. The half-time of transfer is different for each lipid and is invariant over at least a hundredfold range of unlabeled acceptor concentrations, and a tenfold range of probe concentrations, and exhibits first-order kinetics. Since the fluorescence intensity depends linearly on the microscopic concentrations of the probe in the donor, the observed changes in fluorescence reflect transfer to the unlabeled acceptor. When very low concentrations of labeled high-density lipoproteins (HDL) or vesicles are diluted further, the resulting aqueous phase is no longer saturated with the fluorescent lipid. The partition coefficient of the probe between the apolar region of the donor and the aqueous phase requires transfer of the probe into the aqueous phase. Fluorescent lipids partition from HDL or vesicles into water with the same half-time as that observed for transfer between lipoproteins or between vesicles. The latter observation, coupled with invariant exchange rate with concentration and first-order kinetic behavior, is compelling evidence that the rate-limiting step in the transfer is the dissociation into the aqueous solvent.

Transfer rates of PNA between dimyristoylphosphatidylcholine vesicles is pH dependent, 7.4/sec at pH 7.43, and about 50 times slower at pH 2.8 (0.15/sec). Between pH 2.8 and 7.4 at 30 °C in 0.15 M NaCl, observed rates are the arithmetic sum of the rates of transfer of protonated and ionized PNA;

movement of fatty acid from the internal to external side of the single-walled vesicle appears to be faster than transfer between vesicles [61]. The rate of transfer of PNA between dipalmitoylphosphatidylcholine vesicles is 20% slower than transfer between dimyristoylphosphatidylcholine. When the donor and acceptor vesicles contain different phosphatidylcholines, the observed rates of transfer are characteristic of the donor vesicles, and independent of the acceptor composition.

This kinetic mechanism of desorption has been demonstrated in other experimental systems. Transfer of biosynthetic ^{14}C-cholesterol from the membrane of influenza virus to phosphatidylcholine vesicles [62] shows apparent first-order kinetics with a half-time of about 2.5 h, a rate comparable to that reported for exchange of cholesterol between low-density lipoproteins (LDL) and erythrocytes [63, 64]. Fusion between phosphatidylcholine vesicles during cholesterol transfer has been excluded by ^{1}H-NMR [65]. Movement of phosphatidylcholine between vesicles occurs probably by way of a monomolecular species in the aqueous solution [66–68]. A similar mechanism has been demonstrated for monoacylphosphatidylcholine [69]. From either true or micellar solution, this lipid transfers into LDL or HDL across a dialysis membrane. Incubation of isolated rat adipocytes with serum lipoproteins or lymph chylomicrons containing ^{14}C-cholesterol permits uptake of radioactive cholesterol due to exchange without net movement of sterol [70]. The exchange of cholesterol is a physicochemical process, not energy dependent and not mediated by specific lipoprotein receptors. A similar process, stated not to involve fusion of the particle with the membrane, has been implicated for cholesterol transfer from serum lipoproteins to mycoplasma membranes [71].

If a continuous hydrocarbon region is formed by fusion of two lipoproteins, lipids could diffuse from one lipoprotein to another without transit through an aqueous region. Diffusion of lipids over this small distance, would occur in microseconds [72, 73], a time that is 10^3–10^6 times faster than those experimentally determined, and that would be identical for all lipids. The relatively slow time observed for transfer and different rates of transfer for each lipid class indicate a process other than diffusion is rate limiting. The collision-complex mechanism of transfer of the fluorescent lipids between lipoproteins can be excluded on this basis.

Transfer by Fusion of Surface Films

Lipoproteins, as ordinarily isolated from plasma, are stable assemblies that do not fuse in the absence of severe structural perturbations. However, several treatments that produce lipoprotein fusion have been described recently. Heat [74] or guanidinium hydrochloride treatment [75] of HDL releases apolipoprotein and causes fusion of the residual apolipoprotein-depleted HDL, and produces larger spherical lipoproteins. Incubation of HDL with multilamellar phosphatidylcholine liposomes forms complexes with apoA-I which dissociate from HDL. As a consequence, HDL_2 and HDL_3 are transformed into larger lipoproteins by a process thought to involve fusion of apoA-I-depleted HDL particles [76]. Lipolysis of VLDL *in vitro* in the absence of HDL also produces discoidal particles in the HDL density range that contain apoC, phospholipid and cholesterol [77]. In the presence of HDL_3, the action of lipoprotein lipase *in vitro* on VLDL causes conversion of HDL_3 to larger HDL_2 by increasing the apoC, phospholipid and cholesterol content [78]. The mechanism of transfer has not been studied; the transformation presumably proceeds through fusion of VLDL surface film components with HDL_3. Rapid alteration of lipoprotein composition by lipoprotein lipase, as well as a relative abundance of fatty acid [79], monoacylglycerol [79] or monoacylphosphatidylcholine [80], chemical agents that promote membrane fusion [81], are likely required.

The exact mechanism and the sequence of events needed for fusion of two opposing surface films have not been identified; however, present evidence [82–84] suggests that the fusion process involves discrete regions of low-melting, negatively charged lipids. Fusion of pure phospholipid vesicles induced by these lipids has been examined with electron microscopy and differential scanning calorimetry by *Papahadjopoulos et al.* [82, 83]. They conclude that mixing of vesicle components induced by fatty acid or by monoacylphosphatidylcholine is probably due only to an increased rate of molecular diffusion of components, while true fusion of vesicles is induced by Ca^{++} by changing molecular packing and creating new phase boundaries in the lipid surface film. In more complex membrane system, somewhat different requirements for fusion appear to exist. Monoacylphosphatidylcholine is needed for fusion of multilamellar liposomes and nonphagocytic cultures cell lines [85], while monoacylglycerol induces fusion of erythrocytes [79]. The ability of phosphatidylcholine vesicles containing monoacylphosphatidylcholine and octadecylamine to fuse with Ehrlich ascites tumor cells depends on the surface charge of the vesicles [86]; vesicles without monoacylphosphatidylcholine adhere to the cell membrane, but do not fuse. It is not known for these systems whether Ca^{++} are required, or whether, in the presence of membrane proteins [81, 83], the conditions necessary for fusion are created in other ways. Charge-charge interactions apparently are required to bring lipid surfaces into sufficiently close position for fusion, since the energy barrier is approximately 10 kT, much greater than that available through simple Brownian motion, 1.5 kT [82].

Transfer by Specific Carrier Proteins

The role of plasma albumin in transport of fatty acids between tissues has been known for many years although many aspects of fatty acid binding and transport remain controversial [51]. Monoacylglycerol and monoacylphosphatidylcholine also bind to albumin. In some respects, albumin is comparable to the lipid exchange proteins derived primarily from cell homogenates [47–50]. It is clearly established that these exchange proteins bind equimolar amounts of lipid and have considerable specificity for different lipids. However, they appear only to promote exchange reactions in which there is no net mass transfer. Albumin differs in that it can effect net transfer of fatty acid between tissues. In normal metabolic states, the molar ratio of fatty acid to albumin in plasma is 1; in diseased states, the ratio may increase to 4 [51]. The exact mechanism by which carrier proteins promote lipid exchange, or albumin transfers fatty acid to and from cell membranes, remain to be described. There may be initially penetration of the surface film or membrane by the protein, then a specific lipid-protein association and finally desorption of the complex.

Transfer of Fatty Acid and Monoacylglycerol from the
Chylomicron Surface

It is not possible as yet to describe with certainty the sequence and kinetics of the events by which fatty acid and monoacylglycerol are transferred from chylomicrons. Interaction of the chylomicron with the endothelial cell surface undoubtedly involves protein-protein interactions, and probably also protein-glycosaminoglycan interactions, to bring the chylomicron and endothelial cell plasma membrane surfaces into close apposition. If the separation distance is sufficiently small, fusion of the chylomicron surface film with the external leaflet of the endothelial cell plasma membrane could be induced by high concentrations of fatty acid, monoacylglycerol and monoacylphosphatidylcholine, resulting from hydrolysis of triacylglycerol and phosphatidylcholine. As depicted schematically in figure 2, this nonenzymatic process would provide an interfacial continuum between the external leaflet of the endothelial cell membrane and the surface film of the chylomicron. Thus, products of lipolysis would become components of the external surface of the endothelial cell membrane by lateral movement in the interface, due to the concentration gradient in the continuum between the site of triacylglycerol hydrolysis and the cell surface. By this mechanism, transfer of fatty acyl groups from triacylglycerol to the endothelial cell surface could be accomplished without movement through the aqueous compartment.

If there were no interfacial continuum between the chylomicron and the endothelium, products of lipolysis – fatty acids, monoacylphosphatidylcholine and monoacylglycerol – would have to desorb as single molecules into the aqueous solution (fig. 2) and transfer by molecular diffusion [23]. Experiments with VLDL and 9-(3-pyrenyl)-nanonoic acid indicate that fatty acids, like other lipids, desorb from the lipoprotein surface as single molecules in solution as the rate limiting step. The transfer between VLDL has a half-time of 50–70 msec, depending on the size of the lipoproteins [87]. The kinetics of PNA transfer between dimyristoylphosphatidylcholine vesicles indicate the pK_a of this fatty acid is 7.05, considerably higher than that of organic acids dissolved in aqueous salt solutions [61]. The pK_a of long-chain fatty acids in egg phosphatidylcholine vesicles and multilayers is 7.25, as determined by NMR from the differences in the chemical shifts of the ionized and protonated forms of either $[1-^{13}C]$-hexadecanoic or $[1-^{13}C]$-octadecanoic acid [88]. The pK_a is not greatly affected by fatty acid concentrations or by bulk ionic strength. Thus, at the lipoprotein surface, the rate of desorption will depend on the balance of protonated and ionized fatty acids at the site of hydrolysis. This ratio in turn will affect (a) the surface pressure, (b) the extent to which fatty acids either form a separate phase in the surface or interact with other surface film components, and (c) the quantity of fatty acid which, as the protonated form, dissolves in the lipid core. Desorption of fatty acids from pure lipid monolayers, although small, increases with increasing surface pressure (fig. 3), whereas desorption of monoacylglycerol from lipid monolayers is negligible even at high surface pressure, 35 dyn/cm. Desorption of both lipids is markedly increased by the presence of albumin in the aqueous subphase [25]. Glycosaminoglycans may also modify the organization of bound water at the chylomicron surface, thereby directly influencing the kinetics of desorption of lipids into the aqueous phase.

After desorption of lipolytic products from the lipoprotein surface, several factors could influence the movement of fatty acids, monoacylphosphatidylcholine and monoacylglycerol through the aqueous space between the chylomicron surface and the endothelial cell membrane. The relatively small volume of unstirred solution between the lipoprotein and cell surface

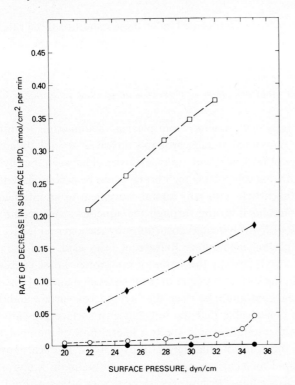

Fig. 3. Effect of albumin on the rate of desorption of oleic acid and monooleoylglycerol. Pure films of either oleic acid (○, □) or monooleoylglycerol (●, ✦) were spread at the interface between argon and an aqueous buffer solution containing, 100 mM Tris-Cl, 1 mM $CaCl_2$ and 0,1 mM EDTA, pH 7.4 containing 4.4 μM (□) or 17.6 μM (✦) albumin. Desorption rates were calculated from the reduction in film area at given pressures. Surface pressure was maintained constant with use of a surface barostat [25].

will limit the number of albumin molecules that can be accommodated in this space. Further, the glycoprotein components of the chylomicron and the glycosaminoglycans thought to be present on the endothelial cell plasma membrane may also restrict diffusion of albumin molecules. A possible role of albumin as a carrier protein in this region does not seem likely since, at low molar ratios of fatty acids to albumin, the high affinity of fatty acids for albumin [51] would limit transfer of fatty acids to cell membranes in certain tissues, especially adipose and mammary. With rapid production of fatty acids by lipoprotein lipase, the high concentration of fatty acid would easily saturate the small aqueous volume and the small number of albumin molecules present in this microenvironment. Under these conditions, the rate of transfer would again be limited by the rate of lipid desorption from the surface.

Transfer of Fatty Acid and Monoacylglycerol across the Endothelial Cell

Net transfer of fatty acyl moieties from plasma chylomicrons to intracellular sites in the individual tissues probably involves several different mechanisms. Transfer of fatty acid and monoacylglycerol between the chylomicron and the endothelial cell plasma membrane could be accomplished as described earlier in this article, either by lateral movement in an interfacial continuum or by molecular diffusion through the aqueous space requiring desorption at the site of hydrolysis and absorption by the external leaflet of the endothelial cell plasma membrane. Release of fatty acids and monoacylglycerol from triacylglycerol by the action of lipoprotein lipase causes a substantial increase in the surface pressure of the interfacial plane [25] (fig. 4). If there is an interfacial continuum between the chylomicron surface and the endothelial cell, it is conceivable that the imbalance in surface pressure at different regions on the interface would create a lateral flow of the lipolytic products away from the point of triacylglycerol hydrolysis. Physical movement of individual lipid molecules is not necessary to maintain a uniform surface pressure in surface films. If, however, there is very little free lipid in the membranes, i.e. lipid molecules not immobilized as boundary lipids associated with proteins, lateral flow of fatty acid and monoacylglycerol might occur, comparable to the spreading of lipid on a lipid-free aqueous surface [89]. The much greater surface area of the endothelial cell, compared to the area around the apolipoprotein:enzyme complex, would create a concentration gradient and thus provide another, perhaps principal, thermodynamic driving force for a unidirectional flux of fatty acid and monoacylglycerol into the cell surface. This effect is found when trioleoylglycerol, applied in excess to the surface of aqueous medium in a monolayer tray with three interconnected compartments, is hydrolyzed by lipoprotein lipase to oleic acid and monooleoylglycerol [25]. The amphipathic lipolytic products immediately locate and spread throughout the interface. As they accumulate in the interface, the surface pressure increases to reduce the rate of triacylglycerol hydrolysis. Simultaneously, substances with lower spreading pressure are displaced from the interface. Addition of albumin to the aqueous medium increases markedly the desorption of oleic acid and monooleoylglycerol from the interface, thereby enhancing lipolysis. Lateral movement

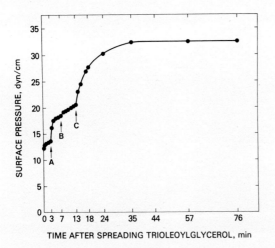

Fig. 4. Increase in surface pressure induced by the action of lipoprotein lipase on trioleoylglycerol spread at the interface. The amount of trioleoylglycerol applied was 14 times that needed to cover the aqueous medium in a monolayer tray with three interconnected compartments with a monolayer at 13 dyn/cm. Additions to the stirred subphase were made at the times indicated by the arrows: A = 60 nM albumin to stabilize the enzyme; B = 26 nM apoC as a source of apoC-II and C = 7 nM lipoprotein lipase. Surface pressure was measured and recorded automatically with a Wilhelmy plate [25].

of lipolytic product in the interface can be observed when enzyme and albumin are added to separate compartments.

A system of spherical vesicles is a prominent feature of capillary endothelium [18, 90, 91] (fig. 5, 6). These structures are very uniform in diameter (60–70 nm) and so numerous as to occupy as much as one third of the cell volume in capillaries of rat myocardium [92]. In cross-section, *Gabbiani and Majno* [93] note that the majority appear just below the plasma membrane, along both the luminal and basal sides of the cell. Some vesicles appear free in the cytoplasm, some have contact with the plasma membrane at the surface of the cell, and others actually open through this membrane, either towards the vascular lumen or towards the basement membrane (fig. 6). The first descriptions of endothelial vesicles suggested that these images might represent different stages of molecular transport through the cytoplasm [90]. More recent studies indicate that vesicles and also vacuoles from channels cross the capillary endothelium [18, 94, 95]. This tubular network of transendothelial channels is continuous with the extracellular space and thus provides a route for movement of lipolytic product across the endothelium. The pres-

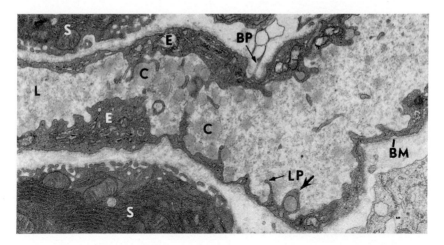

Fig.5. Longitudinal section of a capillary in lactating rat mammary gland taken 10 min after intravenous injection of chylomicrons. Many finger-like processes (LP) are present along the luminal surface of the endothelium (E). Chylomicrons (C) in the lumen (L) are enmeshed by the endothelial processes and partially enveloped by the endothelium (arrows). A slender process of the endothelial cell (BP), associated with irregular lamellar structures, extends from the basal surface toward an extracapillary cell. S = Secretory epithelial cells; BM = basement membrane. × 15,000. From *Scow et al.* [18].

ence of fatty acid in the transendothelial channels during uptake of triacylglycerol from chylomicrons has been observed in tissues incubated after fixation with glutaraldehyde [18,95]. Lipoprotein lipase hydrolyzes chylomicron triacylglycerol in glutaraldehyde-fixed specimens of adipose tissue perfused with chylomicrons when incubated at 38 °C. Morphological studies of specimens treated with Pb^{++} and postfixed with OsO_4 show lamellar and granular structures within vacuoles and vesicles of capillary endothelium and in the subendothelial space between the endothelium and pericytes, but not in the capillary lumen or in or near the fat cells [95]. The lamellar and granular structures, presumably formed by reaction of Pb^{++} with fatty acids, are not present in tissue incubated at 0 °C or in tissues in which lipoprotein lipase activity is inhibited. Similar reaction products are found in transendothelial channels within capillaries of lactating mammary gland [18].

Regardless of how fatty acid and monoacylglycerol are transferred from the chylomicron surface, they will equilibrate spontaneously between the external leaflet of endothelial cell membranes and the extracellular space. It is well established in a variety of model systems that lipids diffuse very rapidly in

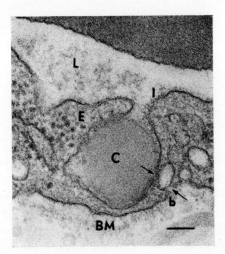

Fig.6. Detail of capillary endothelium in lactating 4-day rat mammary gland taken 10 min after intravenous injection of chylomicrons. A chylomicron (C) is partially enveloped by an endothelial cell. Note that the vesicle in contact with the chylomicron (arrow) spans the width of the endothelium (E) between the chylomicron and the basal (b) surface of the cell. BM = Basement membrane; l = luminal surface of the cell; L = lumen. × 112,000. From *Scow et al.* [18].

the plane of the interface [16, 17]. The lifetime among neighbors for fatty acid in monomolecular films [96], in dipalmitoylphosphatidylcholine vesicles [97] and in liver microsomes [98] is approximately 0.1 μsec. A self-diffusion coefficient of 10^{-8} cm^2 sec^{-1} corresponds to an average travel distance of 1 nm/msec [99].

Since molecules diffuse about 1,000 times more rapidly in aqueous solution than in membranes, molecular diffusion through the aqueous phase could compete with lateral movement in membranes of the transendothelial channels as the primary mechanism of transfer of fatty acid and monoacylglycerol across the endothelial cell. However, data obtained with small-molecular-weight spin labels that partition between the aqueous phase and cellular membranes suggest that the aqueous regions, in cells with extensive membrane structure, have such high microviscosity that diffusion along the plane of the membrane may be faster and more efficient than movement through the three-dimensional aqueous intracellular space [100, 101]. Because the partition coefficient of fatty acids between the aqueous phase and the hydrophobic membrane is about 10^5–10^7 [102–105], we postulate that fatty acids are confined mostly to the membranes of the transendothelial channels, and that the major flux of fatty acid across the endothelial cell occurs by lateral movement within membranes, even though the volume of the membranes is much less than that of the enclosed channels.

Intercellular Transfer of Fatty Acid and Monoacylglycerol

Transfer of fatty acid and monoacylglycerol between cells quite likely involves the same mechanisms as those involved in their transfer from the chylomicron to the plasma membrane of the endothelial cell. As we have postulated earlier [18, 19], fusion of the external leaflet of adjacent cellular plasma membranes would provide an interfacial continuum for lateral movement between cells. This type of fusion occurs between similar cells, as in epithelium, through tight junctions in which only outer leaflets of juxtaposed cell membranes are continuous [106–109]. Lamellar structures, similar to those seen in capillary endothelium by electron microscopy, have been found in spaces between cells, sometimes associated with cellular projections in glutaraldehyde-fixed mammary tissue of lactating rats [19] (fig. 5). When such tissue is incubated, lamellar structures are found closer to secretory epithelial cells, indicating movement of lipolytic product as the result of incubation [19]. The continuity of plasma membrane with layers of lamellar structure suggests that the lamellar are formed as lateral extensions of the external leaflet of membranes during incubation. This interpretation is supported by morphological studies of chylomicrons incubated with lipoproteins lipase in the presence of small amounts of albumin [110]. Lipolytic products accumulate in the interface between triacylglycerol and the aqueous phase and extend the surface film of the chylomicron as a monolayer folded on itself. We propose that lipolytic products move from the site of lipolysis in fixed tissue by extending the interfacial continuum and that lipolysis will continue as long as the interface can expand. Thus, the lamellar structures in fixed tissues mark the route of transit of lipolytic products from chylomicrons, across the capillary endothelium, to cells.

As a competing process, fatty acid and monoacylglycerol could desorb from the basal plasma membrane of the endothelium and diffuse as individual molecules across the intercellular aqueous space. This, of course, would also involve transfer across basement membranes. If fatty acid do desorb in significant amounts, albumin in the extravascular space [111, 112] could act as a trap to limit transfer by this route, depending on the ratio of fatty acid to albumin. As noted previously, the high affinity of albumin for fatty acid would preclude a carrier role of albumin at low fatty acid concentrations; whereas, high-affinity binding sites on albumin would be saturated at high concentrations and desorption of fatty acid from the basal cell surface would, again, be limiting. Monoacylglycerol is hydrolyzed by a specific hydrolase in adipose tissue [113, 114], releasing glycerol into the circulation and fatty acid for uptake by the tissue.

Cellular Uptake and Utilization of Fatty Acids and Monoacylglycerol

Molecular details of the processes for intracellular transport and utilization of fatty acid and monoacylglycerol also remain to be described. After fatty acid and monoacylglycerol become components of the external leaflet of the plasma membrane of parenchymal cells, by mechanisms described above, we suggest that they may move laterally to the external leaflet of the endoplasmic reticulum, which is continuous with that of the plasma membrane [115–117]. Enzymes required for triacylglycerol synthesis are known to be microsomal enzymes [118–120] and reported to be located mainly on the inner surface of the smooth endoplasmic reticulum [121]. There may be a transmembrane multienzyme complex composed of acyl-CoA synthetase [122] and other enzymes required for synthesis of triacylglycerol as integral proteins in the internal leaflet. Formation of fatty acyl-CoA esters and other chemical forms in the cell provides direction for the overall transfer. Triacylglycerol synthesis, which involves sequential reactions of intermediates confined to the membrane, reverses the physical process of phase transfer resulting from lipoprotein lipase action at the chylomicron surface. When diacylglycerol is acylated, the product, triacylglycerol, will separate from the interface in the endoplasmic reticulum to form lenses or droplets between the leaflets of membrane. Formation of the separate triacylglycerol phase also contributes significantly to the thermodynamic driving force for net movement of the fatty acyl moiety into tissue.

Control of Triacylglycerol Hydrolysis

When apoC-II activation of lipoprotein lipase is prevented by high concentrations of anions, lipoprotein lipase remains bound to the lipoprotein substrate and triacylglycerol hydrolysis proceeds, but at a greatly reduced rate [33]. In the metabolic context, we suggest that the important physiological inhibitor of lipoprotein lipase is the fatty acid anion. As a regulatory mechanism, we propose that the concentration of fatty acid in the surface film at the site of triacylglycerol hydrolysis exerts the primary control of the rate and extent of ester bond cleavage in two ways: through product inhibition of the enzyme and disruption of the apoC-II:lipoprotein lipase complex. Lateral separation of apoC-II and lipoprotein lipase in the surface of the lipoprotein would greatly reduce the rate of triacylglycerol hydrolysis, and prevent further production of additional fatty acid.

Movement of fatty acid and monoacylglycerol into tissue for utilization depends on the concentration gradient created when the products are converted in tissue to an apolar chemical form, triacylglycerol, in a separate phase. If fatty acyl-CoA is not utilized, product inhibition of fatty acyl-CoA synthetase will increase the concentrations of both fatty acid and monoacylglycerol throughout the membranes and in the aqueous regions with which they equilibrate. Without a thermodynamic driving force, lipolytic products will not move from the site of lipoprotein lipase action.

Evidence that the rate of fatty acid release by lipoprotein lipase from chylomicrons depends on availability of fatty acid acceptor [123] is presented in figure 7. Rapid release of fatty acid occurs until the molar ratio of fatty acid to albumin is 4. Fatty acid release then continues at a slower rate until the ratio reaches 7; and then slows down further to approximately 1% of the initial rate. As noted previously [110], lipolytic products accumulate *in vitro* in the surface film of the chylomicron when there are not sufficient binding sites on albumin. Lateral extension of the surface film by accumulation of products in the interface allows lipolysis to continue at a slow rate in the absence of acceptors. The inhibition of lipolysis is reversible. If limiting amounts of albumin are added to chylomicrons and lipoprotein lipase, there is an initial rapid release of fatty acid, followed by a much slower release, comparable to that observed when the ratio of fatty acid to albumin exceeds 7 (fig. 7). If, after 60 min of incubation, sufficient albumin is added to accept all potential fatty acids, the rate of lipolysis accelerates at once to that observed initially.

Since triacylglycerol content and the ability of chylomicrons to interact with the vascular bed are not related [124], we propose that products of lipoprotein lipase action on chylomicrons regulate triacylglycerol uptake by tissue at the stage of lipoprotein lipase action. When the tissue can no longer transform the lipolytic products to other chemical forms, the concentration of these substances at the endothelial cell surface will increase. They may act

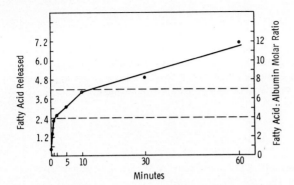

Fig. 7. Dependence of the rate of fatty acid release on the number of available binding sites on albumin. The experimental conditions for hydrolysis of chylomicron triacylglycerol by highly purified lipoprotein lipase have been described [123]. Data of *Scow and Olivecrona* [123, table I], have been plotted.

in two different ways: to prevent the activation of lipoprotein lipase by apoC-II and to act as inhibitory products of the enzymic reaction, thereby greatly reducing or preventing entirely further hydrolysis of triacylglycerol. These mechanisms would provide an immediate response to control triacylglycerol uptake and complement the nutritional and hormonal regulation of the amount of lipoprotein lipase in individual tissues [3, 4].

Conclusions

Transport of fatty acids from chylomicrons in blood to fat cells of adipose tissue requires at least three processes. Chylomicron triacylglycerol is converted by lipoprotein lipase at the surface of capillary endothelium to other acyl lipids, principally fatty acid and monoacylglycerol. This enzymic action allows fatty acyl groups to transfer from the apolar core phase to the interfacial phase of the chylomicron. These lipids then locate and transfer by lateral movement in a continuous interface of various cell membranes to the endoplasmic reticulum of fat cells where they are esterified to triacylglycerol, separate from the interface and accumulate between the bilayers of the endoplasmic reticulum. The separation of triacylglycerol as an apolar phase in the cell provides a thermodynamic driving force for transfer. Control of the transfer process may occur primarily at the endothelial cell surface where the products of lipolysis prevent further hydrolysis of triacylglycerol by lipoprotein lipase.

Acknowledgements

Research support for L.C.S. was provided by The Robert A. Welch Foundation Q-343, United States Public Health Service grants HL-15468, HL-17269 and LRC contract 71-2176. The thoughtful critiques of Dr. *M.C. Doody* are especially appreciated. The authors also thank Mrs. *Sharon Bonnot* for preparation of the manuscript.

References

1 Jackson, R.L.; Morrisett, J.D., and Gotto, A.M., jr.: Lipoprotein structure and metabolism. Physiol. Rev. 56: 259–316 (1976).
2 Herbert, P.N.; Gotto, A.M., and Frederickson, D.S.: Familial lipoprotein deficiency (abetalipoproteinemia, hypobetalipoproteinemia, and Tangier disease); in Stanbury, Wyngaarden and Fredrickson, The metabolic basis of inherited disease; 4th ed., pp. 546–588 (McGraw-Hill, New York 1978).
3 Robinson, D.S.: The function of the plasma triglycerides in fatty acid transport; in Florkin and Stotz, Comprehensive biochemistry, vol. 18, pp. 51–116 (Elsevier, Amsterdam 1970).
4 Scow, R.O.: Metabolism of chylomicrons in perfused adipose and mammary tissue of the rat. Fed. Proc. Fed. Am. Soc. exp. Biol. 36: 182–185 (1977).
5 Smith, L.C.; Pownall, H.J., and Gotto, A.M., jr.: The plasma lipoproteins: structure and metabolism. A. Rev. Biochem. 47: 751–777 (1978).
6 Gangl, A. and Ockner, R.K.: Progress in gastroenterology; intestinal metabolism of lipids and lipoproteins. Gastroenterology 68: 167–186 (1975).
7 Skipski, V.P.: Lipid composition of lipoproteins in normal and diseased states; in Nelson, Blood lipids and lipoproteins: quantitation, composition, and metabolism, pp. 471–583 (Wiley-Interscience, New York 1972).
8 Sata, T.; Havel, R.J., and Jones, A.L.: Characterization of subfraction of triglyceride-rich lipoproteins separated by gel chromatography from blood plasma of normolipemic and hyperlipemic humans. J. Lipid Res. 13: 757–768 (1972).
9 Fraser, R.: Size and lipid composition of chylomicrons of different Svedberg units of flotation. J. Lipid Res. 11: 60–65 (1970).
10 Blanchette-Mackie, E.J. and Scow, R.O.: Scanning electron microscopic study of chylomicrons incubated with lipoprotein lipase. Anat. Rec. 184: 599–609 (1976).
11 Deckelbaum, R.J.; Tall, A.R., and Small, D.M.: Interaction of cholesterol ester and triglyceride in human plasma very low density lipoprotein. J. Lipid Res. 18: 164–168 (1976).
12 Zilversmit, D.B.: Chylomicrons; in Tria and Scanu, Structural and functional aspects of lipoproteins in living systems, pp. 329–368 (Academic Press, London 1969).
13 Jandacek, R.J.; Webb, M.R., and Mattson, F.H.: Effect of an aqueous phase on the solubility of cholesterol in an oil phase. J. Lipid Res. 18: 203–210 (1977).
14 Grundy, S.M. and Mok, H.Y.I.: Chylomicron clearance in normal and hyperlipidemic man. Metabolism 25: 1225–1239 (1976).
15 Scow, R.O.; Hamosh, M.; Blanchette-Mackie, E.J., and Evans, A.J.: Uptake of blood triglyceride by various tissues. Lipids 7: 497–505 (1972).

16 Lee, A. G.: Functional properties of biological membranes: a physical chemical approach. Prog. Biophys. molec. Biol. *29:* 3–56 (1975).
17 Nicolson, G. L.; Poste, G., and Ji, T. H.: The dynamics of cell membrane organization; in Poste and Nicolson, Dynamic aspects of cell surface organization. Cell surface reviews, vol. 3, pp. 1–74 (North-Holland, Amsterdam 1977).
18 Scow, R. O.; Blanchette-Mackie, E. J., and Smith, L. C.: Role of capillary endothelium in the clearance of chylomicrons. A model for lipid transport from blood by lateral diffusion in cell membranes. Circulation Res. *39:* 149–162 (1977).
19 Scow, R. O.; Blanchette-Mackie, E. J., and Smith, L. C.: Role of lipoprotein lipase and capillary endothelium in the clearance of chylomicrons from blood: a model for lipid transport by lateral diffusion in cell membranes; in Polonovski, Cholesterol metabolism and lipolytic enzymes, pp. 143–164 (Masson, New York 1977).
20 Fielding, C. J. and Havel, R. J.: Lipoprotein lipase. Archs Path. *101:* 225–229 (1977).
21 Eisenberg, S. and Levy, R. I.: Lipoprotein metabolism. Adv. Lipid Res. *13:* 1–89 (1975).
22 Ekman, S. and Lundberg, B.: Phase equilibria and phase properties in systems containing lecithins, triglycerides and water. Acta chem. scand. B *32:* 197–202 (1975).
23 Dietschy, J. M.: General principles governing movement of lipids across biological membranes; in Dietschy, Gotto and Ontko, Disturbances in lipid and lipoprotein metabolism, pp. 1–28 (American Physiological Society, Bethesda 1978).
24 Simmonds, W. J.: Uptake of fatty acid and monoglyceride; in Rommel, Goebell and Böhmer, Lipid absorption: biochemical and clinical aspects, pp. 51–64 (University Park Press, Baltimore 1976).
25 Scow, R. O.; Desnuelle, P., and Verger, R.: Lipolysis and lipid movement in a membrane model (submitted for publication, 1978).
26 Scow, R. O. and Egelrud, T.: Hydrolysis of chylomicron phosphatidylcholine *in vitro* by lipoprotein lipase, phospholipase A_2 and phospholipase C. Biochim. biophys. Acta *431:* 538–549 (1976).
27 Fielding, P. E.; Shore, V. G., and Fielding, C. J.: Lipoprotein lipase: properties of the enzyme isolated from post-heparin plasma. Biochemistry, N. Y. *13:* 4318–4322 (1977).
28 LaRosa, J. C.; Levy, R. I.; Herbert, P.; Lux, S. E., and Fredrickson, D. S.: A specific apoprotein activator for lipoprotein lipase. Biochem. biophys. Res. Commun. *41:* 57–62 (1970).
29 Havel, R. J.; Shore, V. G.; Shore, B., and Bier, D. M.: Role of specific glycopeptides of human serum lipoproteins in the activation of lipoprotein lipase. Circulation Res. *27:* 595–600 (1970).
30 Miller, A. L. and Smith, L. C.: Activation of lipoprotein lipase by apolipoprotein glutamic acid. J. biol. Chem. *248:* 3359–3362 (1973).
31 Chung, J. and Scanu, A. M.: Isolation, molecular properties, and kinetic characterization of lipoprotein lipase from rat heart. J. biol. Chem. *252:* 4202–4209 (1977).
32 Fielding, C. J. and Fielding, P. E.: The activation of lipoprotein lipase by lipase coprotein (apoC-2); in Polonovski, Cholesterol metabolism and lipolytic enzymes, pp. 165–172 (Masson, New York 1977).
33 Fielding, C. J. and Fielding, P. E.: Mechanism of salt-mediated inhibition of lipoprotein lipase. J. Lipid Res. *17:* 248–256 (1976).
34 Kinnunen, P. K. J.; Jackson, R. L.; Smith, L. C.; Gotto, A. M., jr., and Sparrow, J. T.:

Activation of lipoprotein lipase by native and synthetic fragments of human plasma apolipoprotein C-II. Proc. natn. Acad. Sci. USA *74:* 4848–4851 (1977).

35 Morrisett, J.D.; Jackson, R.L., and Gotto, A.M., jr.: Lipid-protein interactions in the plasma lipoproteins. Biochim. biophys. Acta *472:* 93–133 (1977).

36 Sparrow, J.T.: unpublished.

37 Doody, M.C.; Kinnunen, P.K.J., and Smith, L.C.: unpublished.

38 Segrest, J.P.; Jackson, R.L.; Morrisett, J.D., and Gotto, A.M., jr.: A molecular theory of lipid-protein interactions in the plasma lipoproteins. FEBS Lett. *38:* 247–253 (1974).

39 Hensok, L.C. and Schotz, M.C.: Detection and partial characterization of lipoprotein lipase in bovine aorta. Biochim. biophys. Acta *409:* 360–366 (1975).

40 Dicorleto, P.E. and Zilversmit, D.B.: Lipoprotein lipase activity in bovine aorta. Proc. Soc. exp. Biol. Med. *148:* 1101–1105 (1975).

41 Olivecrona, T.; Bengtsson, G.; Marklund, S.E.; Lindahl, U., and Höök, M.: Heparin-lipoprotein lipase interactions. Fed. Proc. Fed. Am. Soc. exp. Biol. *36:* 60–65 (1977).

42 Brown, M.S. and Goldstein, J.L.: Receptor-mediated control of cholesterol metabolism. Study of human mutants has disclosed how cells regulate a substance that is both vital and lethal. Science *191:* 150–154 (1976).

43 Mahley, R.W. and Innerarity, T.L.: Interaction of canine and swine lipoproteins with the low density lipoprotein receptor of fibroblasts as correlated with heparin/manganese precipitability. J. biol. Chem. *252:* 3980–3986 (1977).

44 Vandenheuvel, F.A.: Lipid-protein interactions and cohesional forces in the lipoprotein systems of membranes. J. Am. Oil Chem. Soc. *43:* 258–264 (1966).

45 Gurd, F.R.N.: Some naturally occurring lipoprotein systems; in Hanahan, Lipid chemistry, pp. 260–325 (Wiley, New York 1960).

46 Bruckdorfer, K.R. and Graham, J.M.: The exchange of cholesterol and phospholipids between cell membranes and lipoproteins; in Chapman and Wallach, Biological membranes, vol. 3, pp. 103–152 (Academic Press, London 1976).

47 Kader, J.C.: Exchange of phospholipids between membranes; in Poste and Nicolson, Dynamic aspects of cell surface organization. Cell surface reviews, vol. 3, pp. 127–204 (North-Holland, Amsterdam 1977).

48 Zilversmit, D.B. and Hughes, M.E.: Phospholipid exchange between membranes; in Korn, Methods in membrane biology, vol. 7, pp. 211–259 (Plenum Press, New York 1976).

49 Wirtz, K.W.A.: Transfer of phospholipids between membranes. Biochim. biophys. Acta *344:* 95–117 (1974).

50 Ockner, R.K. and Manning, J.A.: Fatty acid-binding protein in small intestine. Identification, isolation and evidence for its role in cellular fatty acid transport. J. clin. Invest. *54:* 326–338 (1974).

51 Spector, A.: Fatty acid binding to plasma albumin. J. Lipid Res. *16:* 165–179 (1975).

52 Bell, F.P.: Lipoprotein lipid exchange in biological systems; in Day and Levy, Low density lipoproteins, pp. 111–113 (Plenum Press, New York 1976).

53 Rubinstein, B. and Rubinstein, D.: Interrelationships between rat serum very low density and high density lipoproteins. J. Lipid Res. *13:* 317–320 (1972).

54 Illingworth, D.R. and Portman, O.W.: Independence of phospholipid and protein exchange between plasma lipoproteins *in vivo* and *in vitro*. Biochim. biophys. Acta *280:* 281–289 (1972).

55 Birks, J.B.: Photophysics of aromatic molecules, pp. 301–371 (Wiley-Interscience, New York 1970).
56 Pownall, H.J. and Smith, L.C.: Viscosity of the hydrocarbon region of micelles: measurement by excimer fluorescence. J. Am. chem. Soc. *95:* 3136–3140 (1973).
57 Charlton, S.C.; Olson, J.S.; Hong, K.-Y.; Pownall, H.J.; Louie, D.D., and Smith, L.C.: Stopped flow kinetics of pyrene transfer between human high density lipoproteins. J. biol. Chem. *251:* 7952–7955 (1976).
58 Charlton, S.C.; Hong, K.-Y., and Smith, L.C.: Kinetics of rac-1-oleyl-2-[4(3-pyrenyl) butanoyl]-glycerol transfer between high density lipoproteins. Biochemistry, N.Y. *17:* 3304–3309 (1978).
59 Kao, Y.J.; Charlton, S.C., and Smith, L.C.: Cholesterol transfer to high density lipoproteins, Fed. Proc. Fed. Am. Soc. exp. Biol. *56:* 936 (1977).
60 Sengupta, P.; Sackmann, E.; Juhnle, W., and Scholz, J.P.: An optical study of the exchange kinetics of membrane bound molecules. Biochim. biophys. Acta *436:* 869–878 (1976).
61 Doody, M.C.; Pownall, H.J., and Smith, L.C.: Mechanism of fatty acid transfer between phosphatidylcholine single walled vesicles. Fed. Proc. Fed. Am. Soc. exp. Biol. *37:* 1832 (1978).
62 Lenard, J. and Tohman, J.E.: Transbilayer distribution and movement of cholesterol and phospholipid in the membrane of influenza virus. Proc. natn. Acad. Sci. USA *73:* 391–395 (1976).
63 Bjornson, L.K.; Gniewkowski, C., and Kayden, H.J.: Comparison of exchange of α-tocopherol and free cholesterol between rat plasma lipoproteins and erythrocytes. J. Lipid Res. *16:* 39–53 (1975).
64 Basford, J.M.; Glover, J., and Green, C.: Exchange of cholesterol between β-lipoproteins and erythrocytes. Biochim. biophys. Acta *84:* 764–766 (1964).
65 Haran, N. and Shporer, M.: Proton magnetic resonance study of cholesterol transfer between egg yolk lecithin vesicles. Biochim. biophys. Acta *465:* 11–18 (1977).
66 Martin, F.T. and MacDonald, R.C.: Phospholipid exchange between bilayer membrane vesicles. Biochemistry, N.Y. *15:* 321–327 (1976).
67 Duckwitz-Peterlein, G.; Eilenberger, G., and Overath, P.: Phospholipid exchange between bilayer membranes. Biochim. biophys. Acta *469:* 311–325 (1977).
68 Thilo, L.: Kinetics of phospholipid exchange between bilayer membranes. Biochim. biophys. Acta *469:* 326–334 (1977).
69 Portman, O.W. and Illingworth, D.R.: Lysolecithin binding to human and squirrel monkey plasma and tissue components. Biochim. biophys. Acta *326:* 34–42 (1973).
70 Kovanen, P.T. and Nikkila, E.A.: Cholesterol-exchange between fat cells, chylomicrons and plasma lipoproteins. Biochim. biophys. Acta *441:* 357–369 (1976).
71 Sltuzky, G.M.; Razin, S.; Kahane, I., and Eisenberg, S.: Cholesterol transfer from serum lipoproteins to mycoplasma membranes. Biochemistry, N.Y. *16:* 5158–5163 (1977).
72 Lee, A.G.; Birdsall, J.M., and Metcalfe, J.L.: Nuclear magnetic relaxation and the biological membrane; in Korn, Methods in membrane biology, vol. 2, pp. 1–156 (Plenum Press, New York 1974).
73 Jain, M.K. and White, H.B. III: Long-range order in biomembranes. Adv. Lipid. Res. *15:* 1–60 (1977).

74 Tall, A.R. and Small, D.M.: Solubilization of phospholipid membranes by human plasma high density lipoproteins. Nature, Lond. 265: 163–164 (1977).
75 Nichols, A.V.; Gong, E.L.; Blanche, P.J.; Forte, T.M., and Anderson, D.W.: Effects of guanidine hydrochlorides on human plasma high density lipoproteins. Biochim. biophys. Acta 446: 226–239 (1976).
76 Tall, A.R.; Hogan, V.; Askinazi, L., and Small, D.M.: Interaction of plasma high density lipoproteins with dimyristoyllecithin multilamellar liposomes. Biochemistry, N.Y. 17: 322–326 (1978).
77 Chajek, T. and Eisenberg, S.: Very low density lipoprotein. Metabolism of phospholipids, cholesterol, and apolipoprotein C in the isolated perfused rat heart. J. clin. Invest. 62: 1654–1665 (1978).
78 Patsch, J.R.; Gotto, A.M.; Eisenberg, S., and Olivecrona, T.: Formation of high density lipoprotein-like particles during lipolysis of very low density lipoproteins *in vitro*. Proc. natn. Acad. Sci. USA 75: 4519–4523 (1978).
79 Ahkong, Q.F.; Fisher, D.; Tampion, W., and Lucy, J.A.: The fusion of erythrocytes by fatty acids, ester, retinol, and α-tocopherol. Biochem. J. 136: 147–155 (1973).
80 Poznansky, M.J. and Weglicki, W.B.: Lysophospholipid induced volume changes in lysosomes and in lysosomal lipid dispersions. Biochem. biophys. Res. Commun. 58: 1016–1021 (1974).
81 Lucy, J.A.: Role of lipids in membrane fusion; in Dils and Knudsen, Regulation of fatty acid and glycerolipid metabolism, pp. 63–72 (Pergamon Press, New York 1977).
82 Papahadjopoulos, D.; Hui, S.; Vail, W.J., and Poste, G.: Studies on membrane fusion. I. Interactions of pure phospholipid membranes and the effect of myristic acid, lysolecithin, proteins and dimethylsulfoxide. Biochim. biophys. Acta 448: 245–264 (1976).
83 Papahadjopoulos, D.; Vail, W.J.; Pangborn, W.A., and Poste, G.: Studies on membrane fusion. II. Induction of fusion in pure phospholipid membranes by calcium ions and other divalent metals. Biochim. biophys. Acta 448: 265–283 (1976).
84 Poste, G. and Papahadjopoulos, D.: Lipid vesicles as carriers for introducing materials into cultured cells: influence of vesicle lipid composition on mechanisms of vesicle incorporation into cells. Proc. natn. Acad. Sci. USA 73: 1603–1607 (1976).
85 Weissmann, G.; Cohen, C., and Hoffstein, S.: Introduction of enzymes, by means of liposomes, into non-phagocytic human cells *in vitro*. Biochim. biophys. Acta 498: 375–385 (1977).
86 Martin, F.J. and MacDonald, R.C.: Lipid vesicle-cell interactions. II. Induction of cell fusion. J. Cell Biol. 70: 506–514 (1976).
87 Doody, M.C.: unpublished.
88 Egret-Charlier, M.; Sanson, A.; Ptak, M., and Bouloussa, O.: Ionization of fatty acids at the lipid-water interface. FEBS Lett. 89: 313–316 (1978).
89 Gaines, G.L.: Insoluble monolayers at liquid-gas interfaces (Wiley-Interscience, New York 1966).
90 Palade, G.E.: Fine structure of blood capillaries. J. appl. Phys. 24: 1424 (1953).
91 Majno, G.: Ultrastructure of the vascular membrane; in Hamilton and Dow, Handbook of physiology, section 2, vol. II, pp. 2293–2375 (American Physiological Society, Washington 1965).
92 Palade, G.E.: Blood capillaries of the heart and other organs. Circulation 24: 368–384 (1961).

93 Gabbiani, G. and Majno, G.: Fine structure of endothelium; in Kaley and Altiva, Microcirculation, vol. I, pp. 134–144 (University Park Press, Baltimore 1977).
94 Simionescu, N.; Simionescu, M., and Palade, G. E.: Permeability of muscle capillaries to small heme-peptides. Evidence for the existence of patent transendothelial channels. J. Cell Biol. 64: 586–607 (1975).
95 Blanchette-Mackie, E. J. and Scow, R. O.: Lipoprotein lipase activity in adipose tissue perfused with chylomicrons. J. Cell Biol. 51: 1–25 (1972).
96 Blank, M. and Britten, J. S.: Transport properties of condensed monolayers. J. Colloid Sci. 20: 789–800 (1965).
97 Lee, A. G.; Birdsall, N. J. M., and Metcalfe, J. C.: Measurement of fast lateral diffusion of lipids in vesicles and in biological membranes by ^{1}H nuclear-magnetic resonance. Biochemistry, N. Y. 12: 1650–1659 (1973).
98 Steir, A. and Sackman, E.: Labels as enzyme substrates. Heterogeneous lipid distribution in liver microsomal membranes. Biochim. biophys. Acta 311: 400–408 (1973).
99 Träuble, H. and Sackman, E.: Studies of the crystalline-liquid crystalline phase transition of lipid model membrane. III. Structure of a steroid-lecithin system below and above the lipid-phase transition. J. Am. chem. Soc. 94: 4499–4510 (1972).
100 Dix, J. A.; Diamond, J. M., and Kirelson, D.: Translational diffusion coefficient and partition coefficient of a spin-labeled solute in lecithin bilayer membranes. Proc. natn. Acad. Sci. USA 71: 474–478 (1974).
101 Keith, A. D. and Snipes, W.: Viscosity of cellular protoplasma. Science 183: 666–668 (1974).
102 Sallee, V. L.: Apparent monomer activity of saturated fatty acids in micellar bile salt solutions measured by a polyethylene partitioning system. J. Lipid Res. 15: 56–64 (1974).
103 Smith, R. and Tanford, C.: Hydrophobicity of long chain n-alkyl carboxylic acids, as measured by their distribution between heptane and aqueous solutions. Proc. natn. Acad. Sci. USA 70: 289–293 (1973).
104 Goodman, D. S.: The distribution of fatty acids between n-heptane and aqueous phosphate buffer. J. Am. chem. Soc. 80: 3887–3892 (1958).
105 Sklar, L. A.; Hudson, B. S., and Simoni, R. D.: Conjugated polyene fatty acids as fluorescent probes: synthetic phospholipid membrane studies. Biochemistry, N. Y. 16: 819–928 (1977).
106 Farquhar, M. G. and Palade, G. E.: Junctional complexes in various epithelia. J. Cell Biol. 17: 375–412 (1963).
107 Staehelin, L. A.: Structure and function of intercellular junctions. Int. Rev. Cytol. 39: 191–283 (1974).
108 Simionescu, M.; Simionescu, N., and Palade, G. E.: Segmental differentiations of cell junctions in the vascular endothelium. J. Cell Biol. 67: 863–885 (1975).
109 Simionescu, M.; Simionescu, N., and Palade, G. E.: Segmental differentiations of cell junctions in the vascular endothelium, arteries and veins. J. Cell Biol. 68: 705–723 (1976).
110 Blanchette-Mackie, E. J. and Scow, R. O.: Retention of lipolytic products in chylomicrons incubated with lipoprotein lipase: electron microscope study. J. Lipid Res. 17: 57–67 (1976).
111 Kaartinen, M.; Kosunen, T. U., and Mäkelä, O.: Complement and immunoglobulin levels in the serum and thoracic duct lymph of the rat. Eur. J. Immunol. 3: 556–559 (1973).

112 Krishnan, L.; Krishnan, E.C., and Jewell, W.R.: Theoretical treatment of the distribution and degradation of vascular, interstitial and intracellular albumin. J. theor. Biol. 67: 609–623 (1977).
113 Tornquist, H. and Belfrage, P.: Purification and some properties of a monoacylglycerol-hydrolyzing enzyme of rat adipose tissue. J. biol. Chem. 251: 813–819 (1976).
114 Zinder, O.; Mendelson, C.R.; Blanchette-Mackie, E.J., and Scow, R.O.: Lipoprotein lipase and uptake of chylomicron triacylglycerol and cholesterol by perfused rat mammary tissue. Biochim. biophys. Acta 431: 526–537 (1976).
115 Robertson, J.B.: Unit membranes: a review with recent new studies of experimental alterations and a new subunit structure in synaptic membranes; in Locke, Cellular membranes in development, pp. 1–81 (Academic Press, New York 1964).
116 Hurry, S.W.: The microstructure of cells, pp. 1–10 (Houghton Mifflin, Boston 1966).
117 Trump, B.F.: The network of intracellular membranes; in Weissman and Claiborne, Cell membranes. Biochemistry, cell biology and pathology, pp. 123–133 (H.P. Publishing, New York 1975).
118 Coleman, R. and Bell, R.M.: Evidence that biosynthesis of phosphatidylethanolamine, phosphatidylcholine, and triacylglycerol occurs on the cytoplasmic side of microsomal vesicles. J. Cell Biol. 76: 245–252 (1978).
119 Hinton, R.H. and Reid, E.: Enzyme distribution in mammalian membranes; in Jamieson and Robinson, Mammalian cell membranes, vol. 1, pp. 161–197 (Butterworths, London 1976).
120 Novikoff, A.B. and Holtzman, E.: Cells and organelles; 2nd ed., pp. 87–99 (Holt, Rinehard & Winston, New York 1976).
121 Higgins, J.A. and Barrnett, R.J.: Fine structural localization of acyltransferases. The monoglyceride and α-glycerophosphate pathways in intestinal absorptive cells. J. Cell Biol. 50: 102–120 (1971).
122 Jason, C.J.; Polokoff, M.A., and Bell, R.M.: Triacylglycerol synthesis in isolated fat cells. An effect of insulin on microsomal fatty acid coenzyme A ligase activity. J. biol. Chem. 251: 1488–1492 (1976).
123 Scow, R.O. and Olivecrona, T.: Effect of albumin on products formed from chylomicron triacylglycerol by lipoprotein lipase *in vitro*. Biochim. biophys. Acta 487: 472–486 (1977).
124 Fielding, C.J. and Higgins, J.M.: Lipoprotein lipase: comparative properties of the membrane-supported and solubilized enzyme species. Biochemistry, N.Y. 13: 4324–4330 (1974).

Dr. L.C. Smith, Department of Medicine A601, Baylor College of Medicine, Houston, TX 77030 (USA)

Very-Low-Density Lipoprotein Metabolism

Shlomo Eisenberg

Lipid Research Laboratory, Department of Medicine B,
Hadassah University Hospital, Jerusalem

Introduction

Very-low-density lipoproteins (VLDL) are one of the two triglyceride-carrying lipoproteins of the blood plasma. VLDL are usually defined as the transport lipoproteins of triglycerides of endogenous (nondietary) origin and are isolated from chylomicron-free plasma at densities of less than 1.006 g/ml. When so isolated, the VLDL consist of lipoprotein particles varying in diameter between 300 and 800 Å, flotation (S_f) rates of 20–400, and molecular weights of $5–100 \times 10^6$ daltons.

VLDL particles are synthesized and secreted by two tissues: liver and intestine. VLDL isolated from plasma contain also products of chylomicron metabolism of S_f rates less than 400, and therefore are a mixture of at least three lipoproteins: VLDL particles of hepatic origin, VLDL particles of intestinal origin and products of chylomicron metabolism (chylomicron 'remnants'). Theoretically, VLDL may represent two independent species of particles, of hepatic and of intestinal origin. Since, however, it is impossible to separate such species, the VLDL are regarded here as defined above, though the definitions, based on flotation properties or source of triglycerides (dietary and nondietary) may be unsatisfactory.

Triglycerides carried in VLDL are transported to sites of storage (adipose tissue) and utilization (heart, muscle, mammary gland, lung, etc.). The key enzymes in VLDL metabolism are the endothelial-bound lipoprotein lipases, discussed in the first and fourth chapters, pages 5 and 109. Following the interaction of VLDL with the lipases, triglyceride-depleted particles are released into the blood stream. These particles are the major source of the circulatory low-density lipoproteins (LDL). During the course of VLDL delipidation (the so-called interconversion process), the particles change composition, structure and biological properties. Hence, the VLDL spectrum represents a dynamic equilibrium of particles of several different sources and at various stages of their metabolic cycle. It is the aim of this communication to describe the pathways of VLDL metabolism, to relate the function and structure of the lipoprotein and to elucidate the potential failures of VLDL metabolism leading to VLDL hyperlipoproteinemia.

Composition and Structure

In view of the marked heterogeneity of VLDL (S_f 20–400), the composition of VLDL particles is best studied in subfractions prepared either by centrifugation [1–5] or gel filtration [6]. The contribution of triglycerides to the total mass of VLDL particles progressively decreases as the particles become heavier and smaller and the percent contribution of proteins, phospholipids and cholesterol increases. Progressive changes of the molar ratios of unesterified to esterified cholesterol and of lecithin to sphingomyelin have been reported. In most studies, but not all, it was found that these ratios decrease with the increased density of the particles [5; exp. 2]. The apoprotein pattern of the different VLDL subfractions is also different. In our studies [4, 5], we have used gel filtration to separate the apoB and apoC. In four fractions (isolated by centrifugation from the plasma of a patient with type V hyperlipoproteinemia) of median S_f rates of 500, 147, 82 and 41, the mass ratios of apoB to apoC were 0.21, 0.30, 0.45 and 0.92, respectively. Similar results were observed in VLDL fractions of a patient with type IV hyperlipoproteinemia and almost normal triglyceride levels. Essentially identical results were reported by *Kane et al.* [7] using tetramethylurea (TMU) to differentiate apoB (TMU-insoluble) and apoC + apoE (TMU-soluble) proteins in VLDL subfraction prepared either by gel filtration or centrifugation. More recently, seven different VLDL subfractions were isolated by zonal centrifutation [8]. The contribution of apoB (TMU-insoluble proteins) to the total protein mass in each fraction was proportional to the density of the VLDL particles. Whether apoE behaves similar to apoB or similar to apoC is not clear. The study of *Kane et al.* [7], indicated an intermediate behavior, with a slower rate of decrease of the ratio of apoE to apoB as compared to the ratio of apoC to apoB. Metabolic studies also demonstrated that apoE contributes more to the protein mass of small VLDL particles as compared to large particles [9]. Some information is available on the relative mass ratios of individual C proteins in different subfractions. *Kane et al.* [7] reported that the percentage of apoC-I increased and that of apoC-II declined progressively with decreasing particle size. In another study [10], it was reported that the percent contribution of apoC-II to the total C protein mass of VLDL density subfractions decreased with decreasing S_f rates, whereas that of apoC-III$_1$ increased. The maximal difference was observed in plasma of patients with type V hyperlipoproteinemia where the ratio of C-II to C-III$_1$ decreased from about 0.50 to 0.25 in fractions of S_f 100–400 and 20–60, respectively. Similar observations have recently been reported in an abstract form [11]. In subfractions obtained from normolipemic humans, however, the ratio of apoC-II to apoC-III$_1$ is similar and does not change much during lipolysis [*Eisenberg*, unpubl.].

Many of the differences in compositon of VLDL subpopulations are reconciled through structural considerations. When viewed through the electron microscope, VLDL appear as spherical particles with a diameter range of 300–800 Å. Structural models of VLDL follow the general core-lipid model of lipoproteins [12] with two major domains: a core containing apolar constituents and an outer shell, approximately 20 Å wide, containing polar and amphipathic constituents [13]. ApoB, C and E, phospholipids and unesterified cholesterol are located predominantly in the outer shell. Cholesteryl esters are randomly distributed throughout the triglyceride core. Some unesterified cholesterol molecules may also be dissolved in the triglyceride core [14]. Because of simple volume to surface considerations, smaller particles must contain relatively less triglycerides and more proteins, phospholipids and unesterified cholesterol than larger particles. Indeed, data on the composition of polar and apolar constituents in VLDL subfractions are in excellent agreement with the core-lipid model [1, 6, 15]. The *absolute amounts* of either core or surface constituents in VLDL particles, however, must decrease with the decreasing size of the particles. A proper analysis of compositional data therefore should relate the data to the *molecular weight* of individual particles and to the *location* of each constituent at either the core or surface of the particle. Such analyses have very seldom been carried out. In our own experiments, we have calculated the absolute mass contribution of lipids and proteins to the total mass of VLDL particles as derived from median S_f rates [5]. With this analysis, we have observed that a 50% decrease of the median molecular weight was associated with a decrease of 60% of the triglyceride mass, 40% of the cholesteryl esters, 40% of apoC and 30 and 40% of the unesterified cholesterol and phospholipids, respectively. The contribution of apoB to the mass of all different particles, in contrast, remained constant. A detailed analysis of seven VLDL fractions isolated by rate-zonal ultracentrifugation

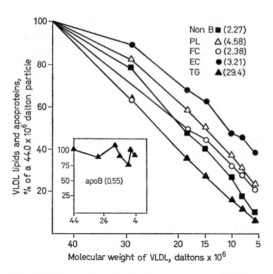

Fig. 1. Lipids and proteins in VLDL particles of different molecular weights. The data are represented as percent of the number of molecules assembled in the largest particles (molecular weight 44.0×10^6 daltons). Numbers in parentheses are the mass contribution (in millions of daltons) of each constituent to the mass of the largest particle. Adapted from *Patsch et al.* [8]. PL = Phospholipid; FC = unesterified cholesterol; EC = cholesterol ester; TG = triglyceride.

[8] is shown in figure 1. Reduced mass contribution of lipid and protein constituents to the total mass of VLDL particles of decreasing molecular weight is obvious, the only exception being apoB.

The concentration of VLDL constituents in the core or at the surface of the lipoprotein has not been systematically studied so far (see, however, 'VLDL Metabolism: Integration of Function and Structure' below). In rat plasma VLDL (average diameter of 427 Å), the combined volume of the core lipids (triglycerides and cholesteryl esters) is in excellent agreement with the calculated core volume [16, 17]. Concentrations of phospholipids and unesterified cholesterol at the surface are 750 and 640 molecules/100,000 Å^2, assuming that both lipids are exclusively localized at the outer shell of the lipoprotein. If the capacity of the triglyceride core to dissolve unesterified cholesterol is about 5 mol% [14], then the concentration of the unesterified cholesterol at the surface decreases to about 510 molecules/100,000 Å^2. Since the molecular weight of rat plasma apolipoproteins is unknown, no such calculations could be carried out for the protein moiety of the VLDL. We have therefore, assigned an arbitrary 'molecular weight unit' of 10,000 daltons to all apoproteins. An average of 55 such units/100,000 Å² area is present at the surface of VLDL. ApoB and apoE contribute 12 'units' each, and apoC the other 31 'units' [17].

Fat Transport in VLDL: the Core

The transport of triglycerides in VLDL occurs through the interaction of the lipoprotein with lipoprotein lipase at the endothelial surface of capillaries in many tissues. It is usually assumed that during a given interaction of a VLDL particle with an enzyme, only a limited amount of triglycerides is hydrolyzed. This assumption is based on the demonstration of product-precursor relationships between the triglyceride [18] and apoB [19, 78] moiety of the circulating small and large VLDL particles. Thus, triglyceride transport in VLDL is viewed as a series of interactions of the particle with lipoprotein lipases at multiple sites. After any given interaction, a particle depleted of some of the triglycerides is released into the blood stream. This particle is then capable of interacting again with the enzyme at another site. After multiple such interactions, particles of S_f rates less than 20 and hydrated density greater than 1.006 g/ml are formed. These particles are operationally defined as intermediate-density lipoproteins (IDL) and are isolated at the salt density interval of 1.006–1.019 g/ml. IDL contain only small amounts of the triglycerides originally present in the VLDL (1–5%); yet, triglycerides contribute about 20% of the mass of IDL particles. This seeming discrepance is explained by loss of other constituents from the VLDL during the interactions of the particle with lipoprotein lipases (*vide infra*). IDL are still regarded as 'triglyceride-rich' lipoproteins and can be further delipidated to LDL. Indeed, precursor-product relationships have been consistently observed between the apoB moiety of VLDL and IDL, and IDL and LDL [5, 20–25]. LDL therefore are the end product of a chain of delipidation steps of the triglyceride-rich VLDL particles.

In analogy to chylomicrons, most of the VLDL triglycerides are probably hydrolyzed by lipoprotein lipases in extrahepatic tissues, predominantly heart, muscle and adipose tissue. To what extent does the hepatic triglyceride lipase participate in the process of VLDL triglyceride transport is unknown. VLDL particles are secreted by the liver and obviously can traverse the hepatic sinusoidal walls and interact with enzymes situated at the hepatocyte plasma membrane. If the hepatic triglyceride lipase is located at the sinusoidal endothelium, then even circulating VLDL particles may interact with the enzyme. Although the purified enzyme is not active towards artificial substrates in the presence of plasma [26, 27], we have recently demonstrated that with VLDL substrates, triglyceride hydrolysis can take place even when the VLDL is dispersed in whole serum [28]. Therefore, the enzyme may – and probably does – hydrolyze some of the VLDL triglycerides. On theoretical grounds, the large and apoC-rich VLDL particles interact preferentially with the apoC-II-activated lipoprotein lipases at extrahepatic sites. As the particle becomes smaller, the number of interactions with the extrahepatic enzymes probably decreases, and the contribution of triglyceride hydrolysis by the hepatic enzyme to the VLDL delipidation process probably increases. According to this hypothesis, the hepatic triglyceride lipase is the predominant enzyme involved with triglyceride hydrolysis at the stage of formation of IDL [29]. Yet, both the hepatic and extrahepatic enzymes can definitely hydrolyze triglycerides from *all* VLDL or IDL particles and LDL-like lipoproteins can be formed *in vitro* when VLDL is incubated with an extrahepatic lipoprotein lipase [30]. Thus, the exact role of the hepatic triglyceride lipase in the overall process of triglyceride transport in VLDL has yet to be established.

The chemical analysis of VLDL and LDL particles has revealed that they contain approximately the same number of apoB molecule(s) in single lipoproteins [5, 29]. Analysis of postlipolysis VLDL, IDL and LDL particles has furthermore revealed that precursor and product particles contain comparable amounts of apoB in single particles [9, 30, 31]. Therefore, *one and only one* product particle is formed from each precursor lipoprotein and neither fission nor fusion of the core of VLDL or IDL takes place during the delipidation process. The structure and composition of postlipolysis VLDL particles produced either *in vitro* [9] or in the supradiaphragmatic portion of the rat [31] have been reported. By electron microscopy, the product lipoprotein is smaller than the precursor and is of spherical shape. It contains less triglycerides and possibly also less cholesteryl esters than the precursor lipoprotein particle. Calculations of the core volume and of the core-lipid volume of postlipolysis particles was carried out in one study [17]; excellent agreement between the two (5.9 and 6.7 million $Å^3$, respectively) was demonstrated. In the two studies, the concentration of polar constituents at the surface of the product lipoprotein was very similar to the precursor. This finding suggests that concomitantly with the decrease of the core volume, surplus surface constituents are removed from the lipoprotein (see 'Fat Transport in VLDL: the Surface' below). However, the apoprotein and possibly also the lipid profile at the outer shell of postlipolysis VLDL particles are very different from the precursor. Some of these changes may be responsible for the decreased fluidity of the lipid domain of postlipolysis VLDL particles as determined by fluorescence depolarization [32].

Triglyceride hydrolysis induces a relative enrichment of the VLDL core with cholesteryl esters. Yet, as shown in figure 1, there is no evidence that small VLDL particles isolated from human plasma contain more cholesteryl ester molecules than larger particles. In our experiments [10, 17], postlipolysis VLDL particles produced during incubation of rat plasma VLDL with lipoprotein-lipase-rich plasma contained *less* cholesteryl ester molecules than the precursor lipoprotein. In the supradiaphragmatic portion of the rat [31], the number of cholesteryl ester molecules increased when 26.6×10^6 dalton 'remnants' were produced from large VLDL particles (molecular weight 56.3×10^6 daltons). These 'remnants' are larger than the intact VLDL used in the *in vitro* experiment (23.0×10^6 daltons). The marked difference in size of the original VLDL used in the two experiments may be responsible for the disparity of the results. According to this view, concomitantly with triglyceride hydrolysis, some cholesteryl esters are transferred to the very large VLDL particles, but not to the average-sized or small particles. Alternatively, the different results may be due to the different experimental conditions. With either interpretation, however, it seems that VLDL particles contain the full complement of cholesteryl ester molecules necessary to form the final delipidation breakdown product, the LDL. With human plasma, they may contain a surplus of cholesteryl esters (fig. 1). In summary, available data suggest that the changing ratio of triglycerides to cholesteryl esters in the different VLDL particles is due predominantly to hydrolysis and removal of triglycerides; some enrichment of the particles with cholesteryl esters cannot, however, be ruled out, especially for the very large, newly secreted particles.

The effects of the changing lipid composition of the VLDL core on the organization and structure of the lipoprotein are probably minimal at densities less than 1.006 or 1.006–1.019 g/ml. In VLDL, the cholesteryl ester molecules are completely dissolved in the liquid triglyceride core of the particle and are not sequestered in a separate domain [33]. Moreover, even in mixtures of VLDL triglycerides and cholesteryl esters containing up to 48% of the weight as cholesteryl esters, no such cooperative cholesteryl ester domains can be identified. From these data, it can be concluded that VLDL does not contain 'LDL' domains within the particle, although all of the LDL constituents are present. The formation of LDL then is due to removal of triglycerides leaving behind the cholesteryl ester moiety together with apoB and surface lipids. It is only when LDL is formed that the typically ordered crystalline state of the cholesteryl esters becomes apparent [34].

Fat Transport in VLDL: the Surface

Apoproteins

ApoC molecules are progressively removed from VLDL during triglyceride hydrolysis [5, 9, 31, 35]. The apoC molecules are preserved in the plasma and are found associated with HDL. They recirculate back to VLDL or chylomicrons when newly synthesized triglyceride-rich lipoproteins enter the circulation [5, 35]. Thus, the circulating apoC mass can be regarded as a discernible pool of apolipoproteins distributed predominantly between VLDL and high-density lipoproteins (HDL). Two different pathways are responsible for the distribution of apoC between VLDL and HDL: exchange and transfer. Exchange of apoC has been demonstrated *in vitro* [36] and immediately after the injection of radiolabeled VLDL or HDL, either to humans [20, 36] or to rats [37–40]. *In vitro* studies have demonstrated unequivocally that the transfer of radioactive apoC from VLDL to HDL is associated with transfer of unlabeled molecules from HDL to VLDL [9, 36]. At equilibrium, the distribution of labeled apoC is proportional to the concentration of the two lipoproteins in the incubation mixture [36], as is the distribution of labeled apoC-II, apoC-III_1 and apoC-III_2 mixed with different human plasma samples [36], and an apoC fraction injected to rats [38]. These results indicate that all of the apoC molecules in VLDL and HDL are readily exchangeable. The possible nature of the exchanging apoC units was studied only very recently [41]. In this study, we have investigated the effects of temperature and plasma on the exchange of apoC between rat plasma VLDL and HDL, and have compared apoC exchange to phospholipids. ApoC exchange occurred rapidly at temperatures of 30 and 37 °C. At the end of 5 min incubation, apoC specific activity in VLDL equaled that in HDL, indicating an equilibrium. At lower temperatures, however, there was a considerable delay in apoC exchange and equilibrium was not reached even after 60 min of incubation. Addition of plasma affected neither the temperature dependence nor the rate of exchange of apoC. Phospholipid exchange was also temperature dependent, but in contrast to apoC, was enhanced ten- to twentyfold by the addition of plasma to the incubation mixture. In the absence of plasma, only minimal exchange of phospholipids between VLDL and HDL was observed even after 120 min incubation. ApoC evidently exchanges between lipoproteins in an unlipidated form or associated with only minimal amounts of phospholipids. This conclusion points to the interesting possibility that the exchange of apoC occurs through a water-soluble form and that at any time a pool of soluble apoC molecules may be present in the water phase.

Transfer of apoC from VLDL to HDL occurs only when VLDL triglycerides are hydrolyzed. It is proportional to the degree of lipolysis [9] and in the presence of plasma, the apoC is found associated with HDL [5, 9, 42]. Hence, it has been suggested that HDL participate in triglyceride transport, being a flexible reservoir of these physiologically important molecules [29, 43]. Yet, VLDL lipolysis is observed in plasma-devoid systems. To study whether the presence of HDL in a lipolytic system is obligatory for the removal of apoC, we have recently determined the fate of VLDL-apoC during lipoprotein-lipase-mediated triglyceride hydrolysis in the absence of plasma or HDL. The experiments were carried out *in vitro*, using purified bovine milk lipoprotein lipase [44, 45] and in the isolated perfused rat heart [46]. In both systems, apoC molecules were removed from the VLDL even in the absence of plasma or HDL, and the rate of removal of apoC was proportional to the rate of triglyceride hydrolysis. The apoC molecules were recovered in the buffer fractions of densities greater than 1.21 g/ml and densities of 1.04–1.21 g/ml, indicating that they are in part poorly lipidated (or even unlipidated) and in part an integral constituent of a lipoprotein fragment of a hydrated density of less than 1.21 g/ml. In a parallel experiment, we have, moreover, observed that the number of apoC molecules removed from VLDL in the absence or presence of plasma was very similar over a range of VLDL triglyceride hydrolysis of 0–50% [44]. Thus, the removal of apoC molecules from VLDL is independent of the presence of HDL in the incubation – or perfusion – mixture; it, therefore, reflects an intrinsic chemical and/or physical change of the VLDL induced by the lipolytic process.

The molecular basis for the dissociation of apoC molecules during VLDL lipolysis is yet unknown. Phospholipids are probably the primary lipid molecules bound to apoproteins in lipoproteins [47]. Since lipoprotein lipases consistently demonstrate phospholipase A activity [48–51], it can be hypothesized that the hydrolysis and removal of phospholipids from VLDL causes dissociation of apoC. However, hydrolysis and removal of as much as 96% of the VLDL glycerophosphatides (by phospholipase A_2), does not cause an appreciable dissociation of either apoC or unesterified cholesterol from the lipoprotein [52]. Thus, the amount of apoC in VLDL particles seems to be best correlated with the size and density of the particle and the amounts of triglycerides. Indeed, triglyceride hydrolysis by another lipase devoid of phospholipase activity (lingual lipase) causes removal of the VLDL-apoC, similar to that observed with lipoprotein lipase [53]. Whether apoC is in actual contact with the triglyceride core of VLDL is unknown.

Neither is the fate of other apoproteins well known. ApoB is apparently fully retained in the particle, though small changes (10–25% of total apoB mass) cannot be ruled out with the presently available techniques. The fate of apoE is not clear. The amount of apoE (isolated by gel filtration) in post-lipolysis VLDL (80% triglyceride hydrolysis) isolated by ultracentrifugation is about 70% of that found in the intact particles [9, 17]. If apoB, C and E are evenly distributed among all VLDL particles, then this result indicates that within the VLDL density range, apoE behaves similar to apoB, i.e. remains associated with the residual core of the lipoprotein. Since there is a very small amount of apoE in LDL, this apoprotein dissociates from VLDL or IDL particles at a very late stage of the delipidation path. Yet, these arguments are not valid if apoE is specifically associated with a subspecies of VLDL and therefore is not removed from the lipoprotein during triglyceride hydrolysis.

Phospholipids

Glycerophosphatides, predominantly lecithin, contribute about 80% to the VLDL phospholipids, the other 15–20% being accounted for by sphingomyelin. Phospholipids exchange among lipoproteins and between lipoproteins and tissue cells. As mentioned earlier in this section, the exchange of lecithin between VLDL and HDL is a very slow process in the absence of plasma. Plasma [41] and the plasma protein fraction with a density greater than 1.21 g/ml [54] enhance the rate of exchange of lecithin ten- to twentyfold. Thus, plasma must contain a specific molecule which enhances the rate of exchange of lecithin between lipoproteins. This molecule is not an apolipoprotein, and is probably analogous to the phospholipid exchange proteins found in liver cell cytosol. Indeed, a protein-containing fraction, prepared from liver cell cytosol, can stimulate lecithin exchange between LDL- and HDL of monkey's serum [55].

Lipoprotein-lipase-mediated lipolysis of VLDL consistently results in transfer of phospholipids from VLDL [5, 9, 31]. As much as 90% of the lecithin molecules present in a VLDL particle of 19.6×10^6 daltons molecular weight are removed concomitantly with triglyceride hydrolysis and the formation of an LDL particle, as are 60% of the VLDL sphingomyelin molecules [5]. One mechanism for removal of lecithin is hydrolysis to lysolecithin, as observed when VLDL is incubated with postheparin plasma obtained from intact rat, the supradiaphragmatic portion of the rat [56] or lipoprotein lipase purified from bovine milk [45, 57]. The maximal amount of lysolecithin formed does not, however, exceed 25–35% of the total lecithin content of the VLDL. By calculation, lysolecithin formation accounts for only one third to one half of the number of lecithin molecules removed from the VLDL. The other half to two thirds must then be removed unhydrolyzed. When plasma is present in the incubation mixture, the unhydrolyzed lecithin molecules removed from the VLDL are found in the HDL density range; lysolecithin is recovered with the plasma protein fraction with a density greater than 1.21 g/ml, presumably bound to albumin [56]. In the absence of plasma, we have observed that the hydrolysis of lecithin to lysolecithin and the removal of unhydrolyzed lecithin molecules progressed at rates comparable to those found when plasma was present [45, 46]. In the two experiments, moreover, the different phospholipids partitioned between density fractions in a manner similar to that observed with plasma, i.e. lysolecithin was found predominantly at a density greater than 1.21 g/ml, and lecithin at densities of 1.04–1.21 g/ml; lecithin distributed also to the fraction with a density greater than 1.21 g/ml.

Sphingomyelin molecules are also removed from VLDL concomitantly with triglyceride hydrolysis. In the plasma-devoid systems, the sphingomyelin is found together with unhydrolyzed lecithin at the fraction with densities of 1.04–1.21 g/ml [45, 46]. At all degrees of triglyceride hydrolysis, and either *in vitro* or during rat heart perfusions, the molar ratio of sphingomyelin to lecithin in the buffer fraction with densities of 1.04–1.21 g/ml was higher than in other fractions, including the VLDL (density smaller than 1.019 g/ml). That fraction (densities of 1.04–1.21 g/ml) was isolated and reincubated with the enzyme. No hydrolysis of lecithin was observed. Similarly, phospholipids in HDL are not a substrate for lipoprotein lipases [58]. We have therefore concluded that the surface of the VLDL particles is the primary site of lecithin hydrolysis and that it occurred during the period of interaction of the lipoprotein with the enzyme. Only after an initial phase of lecithin hydrolysis, does the residual VLDL phospholipid fraction, now enriched with sphingomyelin, dissociate from the VLDL and is found at the buffer fraction with densities of 1.04–1.21 g/ml.

Unesterified Cholesterol

Unesterified cholesterol molecules do exchange between VLDL and HDL and between VLDL and cells, and are being removed from the VLDL during the lipolytic process. The exchange of unesterified cholesterol is temperature dependent but is not enhanced by the presence of plasma proteins [54]. It is thus similar to the exchange of apoC and different from phospholipids.

Initiation of abrupt triglyceride hydrolysis following heparin injection results in increased cholesterol content of human plasma HDL [42], in particular HDL_2 [5, 59]. About 60% of the unesterified cholesterol associated with rat plasma VLDL are removed from the particles subsequent to *in vitro* hydrolysis of 80% of the triglycerides [9, 17]. In the supradiaphragmatic portion of the rat, in contrast, an average decrease of 50% of the weight of large VLDL particles (molecular weight 56.3×10^6 daltons) results in only minimal decrease of the unesterified cholesterol content [31]. These observations suggest that the rate of removal of unesterified cholesterol during lipolysis of VLDL particles of different sizes is different. In the plasma-devoid lipolytic systems [45, 46], unesterified cholesterol is removed from the VLDL throughout the spectrum of triglyceride hydrolysis. The unesterified cholesterol is found almost exclusively in the buffer fraction with densities of 1.04–1.21 g/ml. Since the albumin present in these systems depleted the VLDL of some of the unesterified cholesterol even in the absence of lipolysis, it was impossible to evaluate precisely whether the lipolysis-dependent removal of unesterified cholesterol occurred with a changing or linear rate. It is also unknown whether the distribution of unesterified cholesterol between the core and the surface changes concomitantly with the hydrolysis of triglycerides. Redistribution of cholesterol may cause a shift of unesterified cholesterol to the surface of the lipoprotein that is not reflected by the chemical composition, and may be responsible for the apparently changing rates of removal of unesterified cholesterol from large and small VLDL particles.

VLDL Metabolism: Integration of Function and Structure

On the basis of the data discussed in the previous two sections, it is possible to postulate a scheme of VLDL metabolism based on functional and structural considerations. The main features of the scheme are illustrated in figure 2. The novel hypothesis presented in the scheme is the particulate nature of VLDL metabolism during the process of fat transport. It predicts that subsequent to triglyceride hydrolysis, the VLDL particle disintegrates to form one subunit containing the residual core of the lipoprotein and many subunits containing surplus surface constituents.

VLDL metabolism occurs through multiple interactions with lipoprotein lipases, hydrolysis of triglycerides and decrease of the core volume. During this process, designated the VLDL delipidation path, there is a progressive change of the lipid composition of the core from a predominant triglyceride-rich to a predominant cholesteryl-ester-rich core. The end stage of the delipidation path is the sequestration of LDL where only minuscule amounts

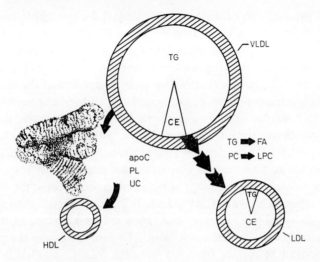

Fig. 2. Schematic representation of VLDL metabolism. TG = Triglyceride; CE = cholesteryl ester; FA = fatty acid; PC = phosphatidylcholine; LPC = lysophosphatidylcholine; PL = phospholipid; UC = unesterified cholesterol.

of the original triglycerides are found (less than 1%). The LDL, however, contain all of the apoB molecules present in the VLDL particle, and most of the cholesteryl esters. This particle shows the typical smectic arrangement of the cholesteryl esters and the characteristic thermal transition over the temperature range of 20–40 °C [60]. The blood plasma, at any given time, contains a spectrum of newly synthesized VLDL particles, partially hydrolyzed particles (including IDL, density 1.006–1.019 g/ml) and LDL particles.

Throughout the VLDL→IDL→LDL delipidation path, surplus surface constituents are removed from the lipoprotein. It is this process that is responsible for the remarkably constant ratio of the surface-to-core constituents in the different VLDL, IDL and LDL. As discussed above ('Fat Transport in VLDL: the Surface'), the process of removal of surface constituents from VLDL is independent of the presence of an acceptor lipoprotein and therefore reflects an intrinsic feature of lipolysis; it is best related to the decrease of the VLDL core volume. In plasma-devoid lipolytic systems, a major part of the surface constituents removed during VLDL lipolysis is found at a lipoprotein density range of 1.04–1.21 g/ml. This finding suggests that they may be associated in a particulate form [30, 45, 46]. Indeed, when viewed through the electron microscope, the density range 1.04–1.21 g/ml buffer fraction contained many disc-shaped structures, similar to those found in the plasma of patients with obstructive jaundice [61], lecithin:cholesterol acyltransferase deficiency [62, 63] and Tangier disease [69]. The experimental conditions do not rule out the possibility that the surface constituents are initially removed in a molecular form, and only then form disc-shaped lipid-protein associations. It is, however, attractive to hypothesize that the discs represent a primary fragment of the VLDL outer shell, and serve as vehicles for the transport of surplus surface constituents from the lipoprotein. According to this hypothesis, the interaction of VLDL with lipoprotein lipase results in hydrolysis of triglycerides and glycerophosphatides at the site of interaction and in the accumulation of surplus surface constituents. The decrease in the core volume and the change of the microenvironment at the site of lipolysis (i.e., presence of hydrolytic products and change of phospholipid composition or phospholipid to cholesterol ratio) induce a change in the physical organization of the surface constituents. This change causes the exclusion of a lipid-protein surface fragment from the VLDL. The 'surface fragments' can then be isolated at a salt density corresponding to their hydrated density. We envision that a very similar process occurs when plasma is present at the site of lipolysis and when the VLDL particle interacts with membrane-bound lipoprotein lipases. In these situations, however, the surface fragments rapidly associate or fuse with HDL particles.

One consequence of triglyceride transport, therefore, is a continuous flux of apoproteins, phospholipids and cholesterol through HDL. It has been recently demonstrated that a major source of the HDL apoproteins (apoA-I and apoA-II) are triglyceride-rich lipoproteins of intestinal origin (p. 200). Lipids and apoC are generated during lipolysis of chylomicrons and VLDL. Moreover, the morphological features of 'surface fragments' generated during lipolysis are very similar to those of HDL precursors ('nascent HDL') isolated from rat liver perfusates [64] or intestinal lymph [65]. Hence, all of the constituents necessary to form HDL are the products of lipolysis and this process may represent the major source of the circulating HDL. It is because of these considerations that a pathway indicating the contribution of surface constituents to HDL has been included in the general scheme of VLDL metabolism (fig. 2). According to this scheme, HDL are in fact a product of the metabolism of the surface coat of chylomicrons and VLDL. HDL thus may represent the final form of the 'surface remnants' whereas LDL are the final form of 'core remnants'. These concepts, while hypothetical, represent a view which regards all plasma lipoproteins as integral blocks of one process, the transport of triglycerides. This view provides a logical explanation for the reciprocal relationships between VLDL and HDL levels [66] and the direct relationships between triglyceride transport (estimated by lipoprotein lipase activity) and HDL [67]. The mechanisms by which 'surface remnants' may contribute – or form HDL are obscure. One mechanism that may transform discoidal HDL precursors to spherical particles is through the lecithin: cholesterol acyltransferase reaction [64]. We suspect that this transformation will be dependent on the availability of apoA from lipolyzed triglyceride-rich lipoproteins of intestinal origin. Another possibility is the incorporation (by fusion ?) of chylomicrons and VLDL surface constituents into preformed HDL particles. This path may transform the smaller HDL_3 particles into the less dense and larger HDL_2. Indeed, such a transformation has recently been demonstrated *in vitro* [68]. Whether some or all of the circulating HDL are products of triglyceride transport is beyond the scope of the present discussion as are the possible molecular events that take place during the formation of HDL.

Let us now turn to the core constituents. An important feature of the hypothesis presented above is the integration of functional and structural considerations in VLDL metabolism. The coordinated decrease of volume and surface results in an almost identical organization of lipids and proteins in precursor and product particles along the VLDL →IDL →LDL path [1, 6, 17, 31]. For example, in our study, the calculated concentration of phospholipids, unesterified cholesterol and protein at the surface of intact and postlipolysis particles yielded almost identical values [17]. Recently, we have been able to carry out similar calculations for seven VLDL fractions isolated by rate-zonal ultracentrifugation [8; fig. 1]. The concentrations of unesterified cholesterol, phospholipids and apoproteins at the surface of VLDL particles of a molecular weight range of $44.0–5.5 \times 10^6$ daltons and a surface area range of $782–181 \times 10^3$ Å^2 were very similar (table I). Yet, it is the alteration of the particles, inherent in the lipolytic process, that is of the greatest biological importance. As described elsewhere in this book, apoC molecules are responsible for fruitful interactions of VLDL with lipoprotein lipases in extrahepatic tissues whereas apoB and apoE are responsible for the interactions with specific receptors at cell surfaces. VLDL metabolism involves major changes of the apoprotein profile of the lipoprotein. It is therefore to be expected that with the progression of the lipolytic process, the affinity of the lipoprotein towards the lipase system and towards cells will change. In figure 3, we have plotted the calculated concentration of apoB and apoC molecular weight 'units' at the surface of seven VLDL particles of a molecular weight range of $44.0–5.5 \times 10^6$ daltons. As is evident from the figure, the change in concentration of either apoB or apoC in different-sized VLDL particles is not linear. There are relatively small changes in apoB concentration when the VLDL molecular weight decreases

Table I. Concentration of unesterified cholesterol, phospholipids and apoprotein units at the surface of seven VLDL density subfractions [calculated from the data of *Patsch et al.,* 8]

flotation rates S_f	Weight daltons $\times 10^6$	Surface area $\text{Å}^2 \times 10^{-3}$	Unesterified cholesterol	Phospholipids	Apoproteins
200	44.0	782	760	780	36.1
139	28.9	586	629	861	38.4
96	18.6	425	692	842	39.3
77	14.8	372	707	828	37.4
50	9.8	275	703	822	37.4
41	8.0	239	694	802	39.7
27	5.5	181	676	795	40.4

Unesterified cholesterol, phospholipids and apoprotein units are expressed as molecules per 100,000 Å^2 surface area. 1 unit = 10,000 daltons.

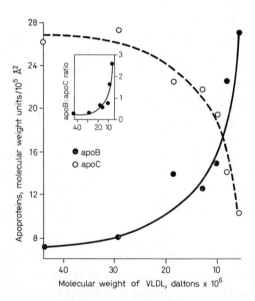

Fig. 3. Concentration of apoB and apoC molecular weight units (10,000 dalton each) at the surface of VLDL particles of different molecular weights. The concentration of apoproteins is represented as units/100,000 Å^2 of surface area. The insert shows the ratio of apoB units to apoC units at the surface of the different particles. Calculated from data of *Patsch et al.* [8].

from 44 to 10×10^6 daltons and when the number of triglyceride molecules per particle decreases from 32,600 to about 6,000. The concentration of apoC decreases during these stages of delipidation only minimally. However, with further delipidation, the concentration of apoB increases, and that of apoC decreases very sharply. An apoB to apoC ratio (fig. 3, insert) best demonstrates the change of the apoprotein profile at the surface of VLDL particles as their molecular weight decreases below 10×10^6 daltons. It is at this value that a sharp change in the biological behavior of the particles is expected. This prediction is in excellent agreement with many observations on VLDL metabolism. In rats, VLDL remnants are cleared rapidly from the circulation prior to the formation of LDL. Both the apoB and cholesteryl ester moieties of the VLDL remnants are found in the rat liver [37, 38, 70, 71], and the clearance of the remnants by the hepatocytes is probably dependent on the interaction of the lipoprotein with a specific, saturable uptake mechanism [72]. A similar path may also operate in the guinea pig [73]. In humans, the capacity of VLDL particles to inhibit 3-hydroxy-3-methylglutaryl-CoA (HMG-CoA) reductase activity in skin fibroblasts becomes apparent only when the particles cross the 10 million dalton molecular weight line [74]. VLDL particles of S_f rates greater than 60 (molecular weight about 11×10^6) do not inhibit the enzyme; a stimulation is often found. The VLDL fraction of S_f rate 20–60, in contrast, has a strong inhibitory effect, indistinguishable from that of LDL. The molecular weight range of these particles is between 5 and 11×10^6 daltons. Since the HMG-CoA reductase inhibitory capacity is present in the large particles and can be demonstrated after *in vitro* lipolysis of the noninhibitory VLDL, the inhibition must represent a change of the noninhibitory VLDL occurring during the delipidation process itself [75]. Still another example is the accumulation of VLDL particles of S_f rates 20–60 following an abrupt initiation of lipolysis by the intravenous injection of heparin [5]. The further delipidation of these particles may be considerably slower due to the decreased concentration of apoC at their surface. All these, as well as other examples of the changing biological behavior of VLDL particles along the delipidation path, are fully explained by the change of the apoprotein pattern at the surface of the particles, as illustrated in figure 4. It should be emphasized, however, that the potential biological importance of other changes of the VLDL outer coat such as increased radius of curvature, decreased fluidity, change of apoE concentration and increased ratio of sphingomyelin to lecithin have not yet been elucidated.

Finally, a remark on the self-regulation of the VLDL metabolism, inherent in the hypothesis is necessary. According to the hypothesis, each of the

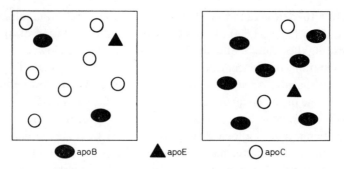

Fig. 4. Graphic representation of apoprotein molecular weight units (10,000 daltons) at a surface area of 25,000 Å² of VLDL particles of molecular weights of 44×10^6 daltons (left) and 5.5×10^6 daltons (right).

many alterations occurring in the VLDL results from the interaction of the particle with lipoprotein lipases and is independent of the presence of other plasma proteins or lipoproteins. The VLDL metabolism may thus be a perfect example for a self-regulatory process, determined solely by the nature of the changing properties of the lipoprotein substrate which, in turn, are determined by the amount of residual triglycerides at each stage of the delipidation path. When operating normally, this process results in fat transport to extrahepatic tissues and formation of LDL and possibly also HDL.

VLDL Metabolism in Humans and Rats

Humans

It was already demonstrated during the 1950s that a protein moiety of human plasma VLDL is a precursor of the protein moiety of LDL [76, 77]. These primary observations were confirmed 15 years later, at a time when methods for the separation and identification of the various VLDL apoproteins became available. Three large studies on VLDL apoprotein metabolism have been published so far [21–23, 25, 78], and several other studies have appeared in abstract or concise forms [19, 24, 79–81]. In all studies, a precursor-product relationship was found between the apoB moiety of VLDL (density smaller than 1.006 g/ml), IDL (density between 1.006 and 1.019 g/ml) and LDL (density between 1.019 and 1.063 g/ml). It also exists between VLDL subfractions of S_f 100–400 (VLDL$_1$), S_f 60–100 (VLDL$_2$), S_f 20–60 (VLDL$_3$), and IDL (S_f 12–20) and LDL_2 (S_f 6–12) [19]. In normal humans, all of the apoB moiety of LDL is derived from VLDL and all of the VLDL-apoB is converted to LDL-apoB [21,78]. The half-life time of apoB in the VLDL density range is between 2 and 4 h, very similar to that reported for triglycerides. A two-compartmental analysis of VLDL-apoB kinetics reveals apoB synthetic rates of about 10 mg/kg/day. In the subjects reported by *Sigurdsson et al.* [21], an excellent agreement was found between VLDL-apoB and LDL-apoB synthetic rates determined by a simultaneous injection of VLDL and LDL, labeled differentially with two isotopes of iodine. Independent catabolism of a fraction of the VLDL-apoB, bypassing LDL formation, however, is necessary to satisfy the observed VLDL-apoB and LDL-apoB kinetics in patients with hypertriglyceridemia, regardless of type [22,23,78]. This direct catabolism of VLDL-apoB may take place in either the VLDL or IDL density range. However, even in those patients whose apoB is cleared from the circulation from the VLDL density range, the clearance occurs after an initial conversion of the large particles ($VLDL_1$ and $VLDL_2$) to small particles ($VLDL_3$) [78]. Thus, although direct catabolism of VLDL apoB may take place in humans, the nature of the particle cleared from the circulation is in agreement with the hypothesis presented in the previous section.

A multicompartmental analysis of VLDL-apoB turnover data has recently been published by *Berman et al.* [25; see also p. 67 this volume]. The validity of the multicompartmental model is strengthened by its similarities to models used for VLDL-triglyceride turnover studies [82,83] and that developed for VLDL-apoC. The analysis simulates a chain of delipidation within the VLDL density range and between VLDL, IDL and LDL. No less than four compartments of delipidation within the VLDL density range are necessary for adequate simulation of the experimental data. The residence time of a particle along the VLDL delipidation path in normal subjects is about 6 h and VLDL-apoB synthetic rates vary between 7 and 11 mg/kg/day. In normal subjects, there is no independent path of clearance of VLDL-apoB or IDL-apoB from the circulation, and no direct synthesis of either IDL-apoB or LDL-apoB.

Several features of VLDL-apoB metabolism are shared by patients with hypertriglyceridemia, regardless of the methodology used in the study or type of hyperlipoproteinemia. In all patients, the rate of VLDL delipidation is considerably slower than in normal subjects and either the half-life or residence time of apoB in the VLDL density range is greatly increased. VLDL-apoB synthetic rates are frequently increased and in most patients there is a direct path of catabolism of the VLDL-apoB bypassing LDL. A very pronounced increase of total apoB synthesis (with either VLDL or IDL) was reported by *Berman et al.* [25] in a group of patients with type III hyperlipoproteinemia. In a more recent study, published in an abstract form, VLDL-apoB synthetic rates in this condition were reported to be almost normal and a delayed conversion of $VLDL_3$ to IDL was assumed to be the major physiological defect of VLDL metabolism in the patients studied [19].

VLDL-apoC metabolism was reported only by the NIH group [25]. An immediate distribution of labeled apoC between VLDL and HDL was observed in all subjects, normals and hyperlipemics. The ratio of labeled apoC in VLDL and HDL was linearly correlated with the mass ratio of the two lipoproteins [36]. ApoC was the major labeled apoprotein in HDL at all times after the injection, and in VLDL after the first 24 h. ApoC is progressively removed from VLDL to HDL particles along the delipidation path and recirculated to newly secreted VLDL particles [25]. The half-life time of apoC in the circulation is about 15–20 h, and it does not vary between normal and hyperlipidemic subjects. ApoC synthetic rates averaged 6 mg/kg/day. The model could not distinguish the sites of entrance and clearance of apoC to and from the circulation and has arbitrarily ascribed them to HDL.

Rats

VLDL-apoB is a precursor of the small amounts of rat plasma LDL-apoB [37, 38, 77], and the apoC moiety of the VLDL exchanges and equilibrates with that present in HDL [37, 38]. The half-life time of apoB in the VLDL density range is very short, 5–10 min. A major difference of VLDL metabolism between rats and humans is the large amounts of apoB cleared directly from the circulation. In two studies [37, 38, 73] using different labeling and analytical techniques, it has been estimated that 85–95% of the VLDL-apoB are cleared from the circulation bypassing LDL. In both studies, the liver was found to be the major site of catabolism of VLDL-apoB from the blood plasma. The fate of VLDL-cholesteryl esters was studied by *Mjøs et al.* [70]. An excellent agreement was found between the fate of VLDL-apoB and that of VLDL-cholesteryl esters. In rats, therefore, the low LDL levels seem to result from an incomplete conversion of VLDL to LDL and the presence of a very efficient mechanism for the clearance of partially lipolyzed VLDL particles from the plasma [37, 38, 73]. As in man, the apoC moiety of rat plasma VLDL equilibrated with that of HDL [38]. Labeled apoC was the major radioactive protein found in either VLDL or HDL at later time intervals (more than 120 min) after the injection of ^{125}I-VLDL to rats [37, 38]. The half-life time of apoC in the rat plasma was 8–10 h, as estimated following the injection of isolated labeled apoC prepared from either VLDL or HDL [38]. Some information was obtained in our study [37, 38] concerning the metabolic fate of VLDL proteins other than apoB and apoC. About 20–30% of the protein-bound radioactivity was associated with apoproteins isolated by gel chromatography on Sephadex G-150 between apoB and apoC. The major apoprotein present in this fraction is apoE. Radioactive protein(s) present with this group were rapidly transferred to HDL and disappeared from the plasma HDL faster than apoA-I or apoC. When isolated and reinjected, this protein distributed among VLDL, HDL and the protein fraction of a density greater than 1.21 g/ml [38] and decayed from the plasma with a half-life time of 7.5 h. Whether the radioactivity followed with this protein(s) is indeed representative of apoE has not been established, however.

The morphological features of the uptake and catabolism of VLDL apoproteins in the rat liver was studied also by us, using light and electron microscopy radioautography [72]. Most of the radioautographic reaction was associated with parenchymal cells. Saturation of the uptake of labeled VLDL by the liver was observed at 5 min after the injection with doses exceeding 1.0 mg protein. During the first 30 min, the radioautographic reaction concentrated at the hepatic cell boundary where it exceeded that of the cytoplasm about fivefold. Cytoplasmic and cell-boundary radioautographic grain density approached one another only 120 min after the injection. At that time interval, many grains were identified in secondary lysosomes. The study thus indicated a saturable mechanism for uptake of VLDL remnants by the rat liver (?receptor-mediated), an initial delay of the particles at the cell boundary and catabolism in secondary lysosomes.

VLDL Hyperlipoproteinemia

Increased levels of lipoproteins result from either increased synthesis or delayed degradation. VLDL are no exception to this rule; however, due to the complexity of VLDL metabolism, regulation of VLDL synthesis and degradation may occur along several different pathway. Since these regulation pathways have been recently discussed in detail [16], they will be only briefly referred to here.

VLDL synthesis is usually equated with either triglyceride synthesis or VLDL-apoB synthesis. Yet, the two are not identical, and it is conceivable that triglyceride transport in VLDL may occur with high, normal or low apoB transport. Theoretically, therefore, an impaired VLDL triglyceride transport may take place with normal plasma triglyceride levels. An example for such a pathway is the secretion of an increased number of small VLDL particles. Plasma VLDL triglyceride levels may be normal, or only slightly elevated, and the main effects will be manifested in the IDL and LDL fractions. Conversely, triglyceride transport in large particles will result in hypertriglyceridemia without an increased number of VLDL particles.

Impairments at stages of VLDL degradation are even more complex. Slow rates of delipidation will result in accumulation of partially degraded VLDL particles. The nature of the VLDL hyperlipoproteinemia will be determined by the delipidation defect and the possible effects of the delayed delipidation on processes such as the type of the newly secreted particles, the rates of interconversion to LDL and the rates of clearance of the partially degraded particles from the plasma. It is, moreover, possible that primary impairment may exist along the interconversion path or at the stage of direct clearance of the partially degraded VLDL from the circulation. Here again, secondary effects on the nature of newly secreted particles and the rates of VLDL delipidation can occur. This situation may explain the observations that several metabolic defects are shared by patients with different forms of VLDL hyperlipoproteinemia (see 'VLDL Metabolism: Integration of Function and Structure' above).

Data presently available do not yet allow us to delineate discrete disorders of fat transport in VLDL. The only exceptions are the rare conditions of low lipoprotein lipase levels [type I hyperlipoproteinemia, 84] and the patient with apoC-II deficiency [85]. The biochemical basis for the elevated levels of VLDL in the majority of patients with VLDL hyperlipoproteinemia is yet unknown. It is hoped, however, that a better understanding of the physiology of fat transport in VLDL and the study of a larger number of patients with VLDL hyperlipoproteinemia will permit discrete disorders to be identified. This, in turn, may lead to a better therapy of VLDL hyperlipoproteinemias.

References

1 Lossow, J.J.; Lindgren, F.T.; Murchio, J.C.; Stevens, G.R., and Jensen, J.C.: Particle size and protein content of six fractions of the $S_f>20$ plasma lipoproteins isolated by density gradient centrifugation. J. Lipid Res. *10:* 68–79 (1969).

2 Hazzard, W.R.; Lindgren, F.T., and Bierman, E.L.: Very low density lipoprotein subfractions in a subject with broad-β disease (type III hyperlipoproteinemia) and a subject with endogenous lipemia (type IV): chemical composition and electrophoretic mobility. Biochim. biophys. Acta *202:* 517–525 (1970).

3   Lindgren, F.T.; Jensen, L.C., and Hatch, F.T.: The isolation and quantitative analysis of serum lipoproteins; in Nelson, Blood lipids and lipoproteins, pp. 181–272 (Wiley, New York 1972).

4   Eisenberg, S.; Bilheimer, D.W.; Lindgren, F.T., and Levy, R.I.: On the apoprotein composition of human plasma very low density lipoprotein subfractions. Biochim. biophys. Acta *260:* 329–333 (1972).

5   Eisenberg, S.; Bilheimer, D.W.; Lindgren, F.T., and Levy, R.I.: On the metabolic conversion of human plasma very low density lipoprotein to low density lipoprotein. Biochim. biophys. Acta *326:* 361–377 (1973).

6   Sata, T.; Havel, R.J., and Jones, A.L.: Characterization of subfractions of triglyceride-rich lipoproteins separated by gel chromatography from blood plasma of normolipemic humans. J. Lipid Res. *13:* 757–768 (1972).

7   Kane, J.P.; Sata, T.; Hamilton, R.L., and Havel, R.J.: Apoprotein composition of very low density lipoprotein of human serum. J. clin. Invest. *56:* 1622–1634 (1975).

8   Patsch, W.; Patsch, J.R.; Kostner, G.M.; Sailer, S., and Braunsteiner, H.: Isolation of subfractions of human very low density lipoproteins by zonal ultracentrifugation. J. biol. Chem *253:* 4911–4915 (1978).

9   Eisenberg, S. and Rachmilewitz, D.: The interaction of rat plasma very low density lipoprotein with lipoprotein lipase rich (post-heparin) plasma. J. Lipid Res. *16:* 451–461 (1975).

10 Carlson, L.A. and Ballantyne, D.: Changing relative proportions of apolipoprotein C-II and C-III with very low density lipoproteins in hypertriglyceridemia. Atherosclerosis *23:* 563–568 (1976).

11 Erkelens, D.V.; Glomset, J.A., and Bierman, E.L.: Preferential binding of apolipoprotein C-II to big triglyceride rich particles. Circulation *55/56:* supp. III, p. 55 (1977).

12 Schneider, H.; Morrod, R.-S.; Colain, J.R., and Tattrie, N.H.: The lipid core model of lipoproteins. Chem. Phys. Lipids *10:* 328–353 (1973).

13 Morrisett, J.D.; Jackson, R.L., and Gotto, A.M.: Lipid-protein interactions in the plasma lipoproteins. Biochim. biophys. Acta *472:* 93–133 (1977).

14 Jandacek, R.J.; Webb, M.R., and Mattson, F.H.; Effect of an aqueous phase on the solubility of cholesterol in an oil phase. J. Lipid. Res. *18:* 203–210 (1977).

15 Shen, B.W.; Scanu, A.M., and Kezdy, F.J.: Structure of human serum lipoproteins inferred from compositional analysis. Proc. natn. Acad. Sci. USA *74:* 837–841 (1977).

16 Eisenberg, S.: Lipoprotein metabolism and hyperlipemia. Atherosclerosis Rev. *1:* 23–60 (1976).

17 Eisenberg, S.: Metabolism of very low density lipoprotein; in Greten, Lipoprotein metabolism, pp. 32–43 (Springer, Heidelberg 1976).

18 Barter, P.J. and Nestel, P.J.: Precursor-product relationship between pools of very low density lipoprotein triglyceride. J. clin. Invest. *51:* 174–180 (1972).

19 Packard, C.J.; Shepherd, J.; Gotto, A.M., and Taunton, O.D.: Kinetic analysis of very low density lipoprotein subfractions in type III and type IV hyperlipoproteinemia. Clin. Res. *25:* 396A (1977).

20 Bilheimer, D.W.; Eisenberg, S., and Levy, R.I.: The metabolism of very low density lipoproteins. I. Preliminary *in vitro* and *in vivo* Observations. Biochim. biophys. Acta *260:* 212–221 (1972).

21 Sigurdsson, G.; Nicoll, A., and Lewis, B.: Conversion of very low density lipoprotein

to low density lipoprotein: a metabolic study of apolipoprotein B kinetics in human subjects. J. clin. Invest. *56:* 1481–1490 (1975).

22 Sigurdsson, G.; Nicoll, A., and Lewis, B.: The metabolism of very low density lipoproteins in hyperlipidemia: studies of apolipoprotein B kinetics in man. Eur. J. clin. Invest. *6:* 167–177 (1976).

23 Sigurdsson, G.; Nicoll, A., and Lewis, B.: The metabolism of low density lipoprotein in endogenous hypertriglyceridemia. Eur. J. clin. Invest. *6:* 151–158 (1976).

24 Phair, R. D.; Hammond, M. G.; Bowden, J. A.; Fried, M.; Fisher, W. R., and Berman, M.: A preliminary model for human lipoprotein metabolism in hyperlipoproteinemia. Fed. Proc. Fed. Am. Socs exp. Biol. *34:* 2263–2270 (1975).

25 Berman, M.; Eisenberg, S.; Hall, M. H.; Levy, R. I.; Bilheimer, D. W.; Phair, R. D., and Goebel, R. H.: Metabolism of apoB and apo C lipoproteins in man: kinetic studies in normals and hyperlipoproteinemics. J. Lipids Res. *19:* 38–56 (1978).

26 LaRosa, J. C.; Levy, R. I.; Windmueller, H. G., and Fredrickson, D. S.: Comparison of the triglyceride lipase of liver, adipose tissue and postheparin plasma. J. Lipid Res. *13:* 356–363 (1972).

27 Chajek, T.; Friedman, G.; Stein, O., and Stein, Y.: Effect of colchicine, cycloheximide and chloroquine on the hepatic triacylglycerol hydrolase in the intact rat and perfused liver. Biochim. biophys. Acta *488:* 270–279 (1977).

28 Eisenberg, S.; Bengtsson, G., and Olivecrona, S.: unpublished observations.

29 Eisenberg, S. and Levy, R. I.: Lipoprotein metabolism; in Paoletti and Kritchevsky, Advances in lipid research, vol. 13, pp. 1–89 (Academic Press, New York 1975).

30 Deckelbaum, R.; Eisenberg, S.; Barenholz, Y., and Olivecrona, T.: Production of human low density lipoprotein-like ('LDL') particles in vitro. J. biol. Chem. *254* (in press, 1979).

31 Mjøs, O.; Faergeman, O.; Hamilton, R. L., and Havel, R. J.: Characterization of remnants produced during the metabolism of triglyceride-rich lipoproteins of blood and plasma and intestinal lymph in the rat. J. clin. Invest. *56:* 603–615 (1975).

32 Barenholz, Y.; Gafni, A., and Eisenberg, S.: Apparent microviscosity of intact and post-lipolysis ('remnant') very low density lipoprotein particles. Chem. Phys. Lipids *21:* 179–185 (1978).

33 Deckelbaum, R. J.; Tall, A. R., and Small, D. M.: Interaction of cholesterol ester and triglyceride in human plasma very low density lipoprotein. J. Lipid. Res. *18:* 164–168 (1977).

34 Deckelbaum, R. J.; Shipley, G. G.; Small, D. M.; Lees, R. S., and George, P. K.: Thermal transitions in human plasma low density lipoprotein Science *190:* 392–394 (1975).

35 Havel, R. J.; Kane, J. P., and Kashyap, M. L.: Interchange of apolipoproteins between chylomicrons and high density lipoproteins during alimentary lipemia in man. J. clin. Invest. *52:* 32–38 (1973).

36 Eisenberg, S.; Bilheimer, D. W., and Levy, R. I.: The metabolism of very low density lipoprotein proteins. II. Studies on the transfer of apoproteins between plasma lipoproteins. Biochim. biophys. Acta *280:* 94–104 (1972).

37 Eisenberg, S. and Rachmilewitz, D.: Metabolism of rat plasma very low density lipoprotein. I. Fate in circulation of the whole lipoprotein. Biochim. biophys. Acta *326:* 378–390 (1973).

38 Eisenberg, S. and Rachmilewitz, D.: Metabolism of rat plasma very low density lipoprotein. II. Fate in circulation of apoprotein subunits. Biochim. biophys. Acta *326:* 391–405 (1973).
39 Eisenberg, S.; Windmueller, H.G., and Levy, R.I.: The metabolic fate of rat and human lipoprotein apoproteins in the rat. J. Lipid Res. *14:* 446–458 (1973).
40 Roheim, P.S.; Hirsch, H.; Edelstein, D., and Rachmilewitz, D.: Metabolism of iodinated high density lipoprotein subunits in the rat. III. Comparison of the removal rate of different subunits from the circulation. Biochim. biophys. Acta *278:* 517–529 (1972).
41 Eisenberg, S.: Effect of temperature and plasma on the exchange of apolipoproteins and phospholipids between rat plasma very low and high density lipoproteins. J. Lipid Res. *19:* 229–236 (1978).
42 LaRosa, J.C.; Levy, R.I.; Brown, W.V., and Fredrickson, D.S.: Changes in high-density lipoprotein composition after heparin induced lipolysis. Am. J. Physiol. *220:* 785–791 (1971).
43 Havel, R.J.: Lipoproteins and lipid transport. Adv. exp. Med. Biol. *63:* 37–59 (1975).
44 Glangeaud, M.C.; Eisenberg, S., and Olivecrona, T.: Very low density lipoprotein. Dissociation of apolipoprotein C during lipoprotein lipase induced lipolysis. Biochim. biophys. Acta *486:* 23–35 (1977).
45 Eisenberg, S. and Olivecrona, T.: J. Lipid Res. (in press, 1979).
46 Chajek, T. and Eisenberg, S.: Very low density lipoprotein. Metabolism of phospholipids, cholesterol, and apolipoprotein C in the isolated perfused rat heart. J. clin. Invest. *61:* 1654–1665 (1978).
47 Segrest, J.P.; Jackson, R.L.; Morrisett, J.D., and Gotto, A.M.: A molecular theory of lipid-protein interactions in the plasma lipoproteins. FEBS Lett. *38:* 247–253 (1974).
48 Vogel, W.C. and Zieve, L.: Post-heparin phospholipase. J. Lipid Res. *5:* 177–183 (1964).
49 Pykalisto, O.J.; Vogel, W.C., and Bierman, E.L.: The tissue distribution of triacylglycerol lipase, monoacylglycerol lipase and phospholipase A in fed and fasted rats. Biochim. biophys. Acta *369:* 254–263 (1974).
50 Ehnholm, C.; Shaw, W.; Greten, H., and Brown, W.V.: Purification from human plasma of a heparin-released lipase with activity against triglycerides and phospholipids. J. biol. Chem. *250:* 6756–6761 (1975).
51 Fielding, P.E.; Shore, V.G., and Fielding, C.J.: Lipoprotein lipase. Isolation and characterization of a second enzyme from postheparin plasma. Biochemistry, N.Y. *16:* 1896–1900 (1977).
52 Eisenberg, S.: Hydrolysis of phosphatidylcholine by phospholipase A_2 does not cause dissociation of apolipoprotein C from rat plasma very low density lipoprotein. Biochim. biophys. Acta *489:* 337–342 (1977).
53 Eisenberg, S.; Blackberg, L., and Olivecrona, T.: unpublished observations.
54 Eisenberg, S.: unpublished observations.
55 Illingworth, D.R. and Portman, O.W.: Independence of phospholipid and protein exchange between plasma lipoproteins *in vivo* and *in vitro*. Biochim. biophys. Acta *280:* 281–289 (1972).
56 Eisenberg, S. and Schurr, D.: Phospholipid removal during degradation of rat plasma very low density lipoprotein *in vitro*. J. Lipid Res. *17:* 578–587 (1976).
57 Scow, R.O. and Egelrud, T.: Hydrolysis of chylomicron phosphatidylcholine *in vitro*

by lipoprotein lipase, phospholipase A_2 and phospholipase C. Biochim. biophys. Acta *431:* 538–549 (1976).
58 Eisenberg, S.; Shurr, D.; Goldmann, H., and Olivecrona, T.: Biochim. biophys. Acta *531:* 344–361 (1978).
59 Nichols, A.V.; Strisower, E.H.; Lindgren, F.T.; Adamson, G.L., and Coggiola, E.L.: Analysis of the change in ultracentrifugal lipoprotein profiles following heparin and ethyl-*p*-chlorophenoxyisobutyrate administration. Clinica chim. Acta *20:* 277–283 (1968).
60 Atkinson, D.; Deckelbaum, R.J.; Small, D.M., and Shipley, G.G.: Structure of human plasma low-density lipoproteins: molecular organization of the central core. Proc. natn. Acad. Sci. USA *74:* 1042–1046 (1977).
61 Hamilton, R.L.; Havel, R.J.; Kane, J.P.; Blaurock, A.E., and Sata, T.: Cholestasis: lamellar structure of the abnormal human serum lipoprotein. Science *172:* 475–478 (1971).
62 Forte, T.; Norum, K.R.; Glomset, J.A., and Nichols, A.V.: Plasma lipoproteins in familial lecithin: cholesterol acyltransferase deficiency: structure of low and high density lipoprotein as revealed by electron microscopy. J. clin. Invest. *50:* 1141–1148 (1971).
63 Norum, K.R.; Glomset, J.A.; Nichols, A.V.; Forte, T.; Albers, J.J.; King, W.C.; Mitchell, C.D.; Applegate, K.R.; Gong, E.L.; Cabana, V., and Gjone, E.: Plasma lipoproteins in familial lecithin: cholesterol acyltransferase deficiency: effect of incubation with lecithin:cholesterol acyltransferase *in vitro*. Scand. J. clin. Lab. Invest. *35:* suppl. 142, pp. 31–55 (1975).
64 Hamilton, R.L.; Williams, M.C.; Fielding, C.J., and Havel, R.J.: Discoidal bilayer structure of nascent high density lipoproteins from perfused rat liver. J. clin. Invest. *58:* 667–680 (1976).
65 Green, P.H.R.; Tall, A.R., and Glickman, R.M.: Rat intestine secretes discoid high density lipoprotein. J. clin. Invest. *61:* 528–534 (1978).
66 Nichols, A.V.: Human serum lipoproteins and their interrelationships. Adv. biol. med. Phys. *11:* 110–158 (1967).
67 Nikkila, E.H.: Metabolic and endocrine control of plasma high density lipoprotein concentration; in Gotto, Miller and Oliver, high density lipoproteins and atherosclerosis (Elsevier/North-Holland, Amsterdam 1978).
68 Patsch, J.R; Gotto, A.M.; Olivecrona, T., and Eisenberg, S.: Formation of high density lipoprotein$_2$-like particles during lipolysis of very low density lipoprotein *in vitro*. Proc. natn. Acad. Sci. USA *75:* 4519–4523 (1978).
69 Assmann, G.; Herbert, P.N.; Fredrickson, D.S., and Forte, T.: Isolation and characterization of an abnormal high density lipoprotein in Tangier disease. J. clin. Invest. *60:* 242–252 (1977).
70 Faergeman, O. and Havel, R.J.: Metabolism of cholesteryl esters of rat very low density lipoproteins. J. clin. Invest. *55:* 1210–1218 (1975).
71 Faegerman, O.; Sata, T.; Kane, J.P., and Havel, R.J.: Metabolism of apo-lipoprotein B of plasma very low density lipoprotein in the rat. J. clin. Invest. *56:* 1396–1403 (1975).
72 Stein, O.; Rachmilewitz, D.; Sanger, L.; Eisenberg, S., and Stein, Y.: Metabolism of iodinated very low density lipoprotein in the rat: autoradiographic localization in the liver. Biochim. biophys. Acta *360:* 205–216 (1974).

73 Barter, P.; Faergeman, O., and Havel, R.J.: Metabolism of cholesteryl Esters of very low density lipoproteins in the guinea pig. Metabolism 26: 615–622 (1977).

74 Gianturco, S.H.; Gotto, A.M., jr.; Jackson, R.L.; Patsch, J.R.; Sybers, H.D.; Taunton, O.D.; Yeshurun, D.L., and Smith, L.C.: Control of 3-hydroxy-3-methyl-glutaryl-CoA reductase activity in cultured human fibroblasts by very low density lipoproteins of subjects with hypertriglyceridemia. J.clin.Invest. 61: 320–328 (1978).

75 Catapano, A.L.; Gianturco, S.H.; Kinnunen, P.K.J.; Eisenberg, S.; Smith, L.C., and Gotto, A.M.: Suppression of HMG-CoA reductase by VLDL remnants made *in vitro* by lipoprotein lipase from non-suppressive VLDL. Fed.Proc. Fed.Am.Socs exp.Biol. 37: 1482 (1978).

76 Volwiler, W.; Goldsworthy, P.D.; MacMartin, M.P.; Wood, P.A.; Mackay, I.R., and Fremont-Smith, K.: Biosynthetic determination with radioactive sulfur of turnover rates of various plasma proteins in normal and cirrhotic Man. J.clin.Invest. 34: 1126–1146 (1955).

77 Gitlin, D.; Cornwell, D.G.; Nakosato, D.; Oncley, J.L.; Hughes, W.G., jr., and Janeway, C.A.: Studies on the metabolism of plasma proteins in the nephrotic syndrome. II. The lipoproteins. J.clin.Invest. 37: 172–184 (1958).

78 Reardon, M.F.; Fidge, N.H., and Nestel, P.J.: Catabolism of very low density lipoprotein B apoprotein in man. J.clin.Invest. 61: 850–860 (1978).

79 Soutar, A.K.; Myant, N.B., and Thompson, G.R.: Simultaneous measurement of apolipoprotein B turnover in very low and low density lipoproteins in familial hypercholesterolemia. Atherosclerosis 28: 247–256 (1977).

80 Ballantyne, F.C.; Ballantyne, D.; Olsson, A.G.; Rossner, S., and Carlson, L.A.: Metabolism of very low density lipoprotein of S_f 100–400 in type V hyperlipoproteinemia. Acta med. scand. 202: 153–161 (1977).

81 Quarfords, S.H.: The turnover of human plasma very low density lipoprotein protein. Biochim. biophys. Acta 489: 477–485 (1977).

82 Shames, D.M.; Frank, A.; Steinberg, D., and Berman, M.: Transport of plasma free fatty acids and triglycerides in man: a theoretical analysis. J.clin.Invest. 49: 2298–2314 (1970).

83 Quarfordt, S.H.; Frank, A.; Shames, D.M.; Berman, M., and Steinberg, D.: Very low density lipoprotein triglyceride transport in type IV hyperlipoproteinemia and the effects of carbohydrate rich diet. J. clin. Invest. 49: 2281–2298 (1970).

84 Krauss, R.M.; Levy, R.I., and Fredrickson, D.S.: Selective measurement of two lipase activities in postheparin plasma from normal subjects and patients with hyperlipoproteinemia. J. clin. Invest. 54: 1107–1125 (1974).

85 Breckenridge, W.C.; Little, J.A.; Steiner, G.; Chow, A., and Poapst, M.: Hyperglyceridemia associated with deficiency of apolipoprotein C-II. New Engl.J.Med. 298: 1265–1273 (1978).

S. Eisenberg, MD, Lipid Research Laboratory, Department of Medicine B, Hadassah University Hospital, Jerusalem (Israel)

Origin, Turnover and Fate of Plasma Low-Density Lipoprotein

D. Steinberg

Department of Medicine, Division of Metabolic Disease, University of California San Diego, La Jolla, Calif.

Introduction

As the pace of lipoprotein research increases and as more powerful techniques of fractionation and characterization are brought to bear, it is becoming apparent that many lipoprotein fractions once thought to be homogeneous are in fact heterogeneous. Isolation by any one technique certainly does not insure homogeneity; successive application of several different techniques (e.g. electrophoresis, preparative ultracentrifugation, gel filtration) increases the probability that a pure fraction will be obtained but it is seldom, if ever, possible to be confident about purity at the 100% level. In some types of studies, such as studies of the metabolic fate of ^{125}I-labeled, high-specific-activity lipoproteins, even a 1% contamination can be important. Thus, it is essential that a precise operational definition always be given of just how the fraction was prepared and how purity was demonstrated. Sufficient detail should be given to allow retrospective reevaluation should new information become available at a later time.

Like other lipoprotein fractions, 'low-density lipoprotein' (LDL) has been differently defined by different investigators at different times. In some of the early reports, this term was used to encompass all the lipoproteins in the density range 1.006–1.063. Now, the fraction with a density between 1.006 and 1.019 (intermediate-density lipoprotein; IDL, LDL$_1$) is generally believed to consist of lipoproteins derived by partial degradation of very-low-density lipoprotein (VLDL; and possibly also chylomicrons). It may contain, in addition to apolipoprotein B (apoB), varying amounts of apoC and apoE and its metabolism differs from that of the fraction with density 1.019–1.063. The latter is the fraction now generally designated LDL (LDL$_2$). The bulk of material isolated in the density range 1.019–1.063 migrates as a single band with β-mobility on electrophoresis, yields a fairly narrow peak in the analytic ultracentrifuge with S_f° 0–12, and shows a single band in immunoelectrophoresis against antibody to whole human plasma. In man, the apoB in the LDL fraction so defined appears to behave kinetically as a single component in most studies but some departures from 'ideal' behavior have been noted that may have a technical basis or may denote intrinsic heterogeneity [1]. In the rat, however, the fraction with a density between 1.019 and 1.063 is metabolically far from homogeneous [2]. The fraction with a density between 1.019 and 1.040 is derived from VLDL but the fraction with a density between 1.040 and 1.063 is secreted directly into the plasma compartment.

Implicit in the accepted definition of LDL is that it contains only a single apoprotein, apoB. Apoproteins other than apoB can be found in the density range 1.019–1.063 under some conditions [3], but the quantities are insufficient to be accounted for by stoichiometric partition among all the LDL particles, immediately implying heterogeneity. The studies of *Eisenberg* [4] suggest that during the degradation of human VLDL there may be generation of fragments at densities comparable to that of LDL and even high-density lipoprotein (HDL). Some apoC was recovered in the fraction with a density between 1.019 and 1.063 [5]. Finally, in cholesterol-fed animals, *Mahley et al.* [6] have shown that significant amounts of an unusual lipoprotein (HDLc) appear in the density range 1.02–1.087. HDLc contains apoC, apoA and apoE but *no* apoB. It is clear that we must keep in mind the possible presence of more than one lipoprotein species in the 'LDL fraction'.

Evidence for *functional heterogeneity* of LDL has been presented by *Edgington et al.* [7]. They have identified an LDL subfraction in the serum of patients with viral hepatitis that inhibits formation of rosettes with T lymphocytes (Rosette Inhibitory Factor). Another biologically active subspecies of LDL was identified that inhibits *in vitro* stimulation of human lymphocytes by lectins (LDL-In). These properties are definitely *not* shared by all LDL molecules.

These introductory remarks are offered as a caveat before we launch into a discussion of 'LDL'. While we recognize that the fraction of density 1.019–1.063 may not be homogeneous, the large bulk of lipoprotein in this fraction behaves kinetically as if it were a single component, and contains predominantly apoB. Finally, and importantly, it is this fraction and this fraction only, that is elevated in the plasma of patients with familial hypercholesterolemia. This genetically determined hyperlipoproteinemia has been shown to be associated with a specific deletion of a specific cell membrane receptor [8]. This genetic evidence provides strong support for the conclusion that the mechanisms regulating the metabolism of LDL, as defined above, are unique and a legitimate area of discourse. Needless to say, there may be significant differences in LDL metabolism from one animal species to another. In what follows, we shall use data on human LDL wherever possible and indicate the studies done in other species.

Structure of LDL: Implications with Respect to Metabolism

A detailed discussion of LDL structure is beyond the scope of this review. We shall be concerned only with a few basic facets, particularly relevant to our discussion of the metabolism of the LDL molecule. Excellent reviews of lipoprotein structure have appeared recently [9, 10] and can be consulted for primary references.

Human LDL is a spherical molecule with diameter in the range of 200–250 Å as determined by electron microscopy. The molecular weight is 2.0–2.5 million. Protein makes up 20–25% of the mass and essentially all of it is apoB. Because apoB once released from LDL (or VLDL) by delipidation is virtually insoluble, it has not been well characterized. Even the unit molecular weight is very much in doubt, reported values ranging from 10,000 to 250,000. The number of peptide fragments obtained by treatment with cyanogen bromide suggests that the true molecular weight may be nearer the lower end of this range [11]. The question of how many identical subunits are present on each particle is highly relevant to considerations of the binding and uptake of the hololipoprotein. For example, if there are 10–20 identical subunits per particle, each particle could bind to multiple receptor sites on the cell membrane. This could have important metabolic implications. For example, the binding affinity would be much greater than it would be if there were only one site of attachment per molecule. With multiple sites of LDL-membrane interaction, the rate of dissociation would be very slow, as it indeed appears to be. In addition, there might be a bridging of multiple receptor sites that could facilitate the internalization process.

ApoB is immunochemically distinct from all of the other apolipoproteins, including apoE. Yet apoE-containing lipoproteins also bind to the high-affinity LDL receptor, possibly with an affinity even higher than that of apoB-containing lipoproteins [12–14]. The implication is that these two apoproteins have at least some important structural features in common (or that the cell receptor includes two recognition sites). The common structural features involved, however, may be only weakly antigenic. Conceivably, apoB may include peptide sequences or carbohydrate sequences similar to or identical with sequences also found in apoE. However, because non-shared sequences are much more dominant antigenically, the antisera against these two apoproteins fail to show cross-reaction. Until more protein chemistry becomes available we can only speculate, but the possibility that apoB and apoE are somehow biogenetically related should be considered. ApoB, like most plasma proteins, is a glycoprotein and the carbohydrate moiety has been partially characterized [15]. It includes terminal sialic acid residues, raising the possibility that removal of LDL from the plasma might be accelerated by desialylation as demonstrated by *Ashwell and Morell* [16] for a number of plasma glycoproteins. These proteins (e.g. ceruloplasmuin, fetuin) are removed predominantly by the liver whereas LDL appears to be removed predominantly by extrahepatic tissues [17, 18]. Still, the liver does have some capacity to degrade LDL [19–21] and that fraction removed by the liver might represent disialylated LDL. Preliminary studies by *Attie et al.* [22], however, fail to show any difference in the turnover of native and enzymatically desialylated human LDL in the pig *in vivo* or in liver cells in culture. Further studies are needed to better characterize the carbohydrate moiety to determine whether it is modified after secretion of LDL and what effect, if any, such a modification has on the metabolism of the lipoprotein.

Cholesterol and cholesterol esters make up about 50% of the weight of LDL, phospholipids (mostly lecithin and sphingomyelin) about 20%, and the triglycerides about 10%. The most widely accepted structural model for LDL, like that for the other lipoproteins, is one in which the nonpolar components (cholesterol esters and triglycerides) make up the central core of the molecule while the polar components (phospholipids and protein) form an outer shell, making the complex water-soluble. If the model is correct, the nonpolar core lipids would not be expected to play a role in LDL-cell interaction. This has been explicitly demonstrated by showing that LDL from which virtually all of the cholesterol and triglyceride have been extracted is metabolized by cultured fibroblasts in a manner not distinguishable from that for intact LDL [23]. The heptane extraction method of *Gustafson* [24] was modified so that a soluble 'cholesterol-free' particle could be obtained in good yield. This modified LDL was bound, internalized and degraded like native LDL. However, because it lacked cholesterol, it was not able to suppress endogenous cholesterol synthesis [8]. The results show that interaction of the lipoprotein with the LDL receptor is a necessary but not a sufficient cause for suppression of 3-hydroxy-3-methyl-glutarylcoenzyme A (HMG CoA) reductase activity. A similar conclusion comes from studies with solubilized apoB (stabilized with albumin) showing normal binding but no inhibition of cholesterol synthesis [25].

The configuration of the apoprotein in native LDL is not known. Presumably changes in that configuration might, if sufficiently radical, affect the affinity of LDL for the cell surface and thus affect LDL turnover. A report has appeared describing a kindred in which the plasma LDL failed to suppress endogenous cholesterol synthesis, implying a structural abnormality [26]. However, *in vivo* studies showed normal turnover when that patient's LDL was injected into a normal recipient [27]. Degradation *in vitro* was also normal both in lymphocytes [27] and in fibroblasts [*J.P.D. Reckless, D.B. Weinstein, D.S. Galton and D. Steinberg*, unpubl.]. *Malloy and Kane* [28] have reported a fascinating case of 'abetalipoproteinemia' in which the apoB found in VLDL shows a molecular weight only about 60% that of normal apoB, perhaps the first described mutation radically affecting apoB structure.

Origin of LDL

Studies in several laboratories have established beyond doubt that plasma VLDL can be converted to LDL in man and in several animal species, presumably through the action of lipoprotein lipase [29–36]. A precursor-product relationship has been shown using triglyceride labeling, cholesterol ester labeling, and protein (apoB) labeling; the precursor-product relationship holds when isolated VLDL is labeled *in vitro* and injected and also when lower-molecular-weight precursors are injected *in vivo* (amino acids, free fatty acids, glycerol). The conversion to LDL occurs in a series of steps (number indeterminate and probably variable) in which the VLDL eventually loses most of its triglyceride, all the apoproteins other than apoB, some phospholipid and some cholesterol (fig. 1, pathways 1, 7–10). Conversion of larger VLDL to smaller VLDL has been demonstrated *in vitro* [37] and *in vivo* [38–40] (fig. 1, pathway 7–8).

But is VLDL the sole precursor of LDL? In some species and under some circumstances, the answer appears to be that it is (at least within the quantitative limitations of the methods used). As indicated above, in normal man the curve for VLDL specific radioactivity intersects that for LDL at its peak, compatible with a direct and exclusive precursor role for VLDL. Moreover, the total flux of LDL apoB (mg/kg/day) in normal man has been reported to match very closely the total flux of apoB in VLDL [35]. Until very recently,

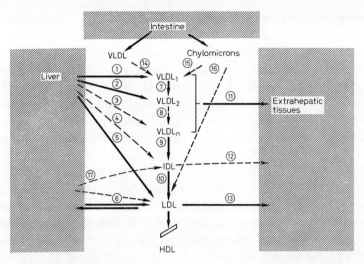

Fig. 1. Schematic representation of the origins of LDL apoB, established (solid arrows) and speculative (dashed arrows). Pathway 6 represents the rapid exchange between plasma LDL and the large pool of LDL in the liver. Direct hepatic secretion of LDL is shown as pathway 5. The extent to which the exchange of LDL into the liver represents potential for irreversible uptake (i.e. apoB degradation) is not known. For discussion of the other pathways and the evidence implicating them, please see the text.

then, it was generally held that there was little or no direct hepatic (or intestinal) secretion of LDL and that *all* LDL was derived from degradation of VLDL. This conclusion is consonant with the finding that the calculated mean apoB content of a VLDL particle is the same as that of an LDL particle [41]. Thus, degradation and structural modification of a VLDL particle (by lipoprotein lipase, hepatic lipase, LCAT and possibly other enzymes) can theoretically generate a particle of the apoprotein and lipid composition characteristic of plasma LDL. Do the final steps in the conversion of VLDL to LDL occur within the plasma compartment by simply continuing the lipoprotein-lipase-catalyzed degradation of VLDL? Or does the liver play a role as suggested by *Felts and Itakura* [42]? They propose that small VLDL or IDL are taken up by or bound to hepatocytes and only then is the final transition to LDL effected (fig. 1, pathway 17). Whether or not hepatic lipase plays a role in VLDL-LDL interconversion remains to be determined.

The first clear evidence that *de novo* secretion of LDL can be quantitatively significant came from studies in patients with homozygous familial hypercholesterolemia. *Soutar et al.* [43] found that LDL apoB flux in these patients was 50–100% greater than their VLDL apoB flux. These results imply that direct secretion of LDL (or IDL) is a major pathway in these patients (fig. 1, pathway 5). *Janus et al.* [44] have confirmed that in patients with hyperlipoproteinemia there can be significant direct secretion of LDL. Using multicompartmental analysis of data obtained after injection of labeled leucine into a diabetic subject with hypertriglyceridemia, *Phair et al.* [40] concluded that as much as 24% of their patient's LDL apoB was directly secreted. *Illingworth* [45] from studies in the normal squirrel monkey, calculated that as much as 14% of the LDL fraction (density range 1.019–1.063) might be directly secreted rather than arising from VLDL. *Nakaya et al.* [46] observed the release of LDL during perfusion of pig liver. Some of this was found to be due to depletion of a pool of performed LDL (see discussion of extravascular LDL pool below), but isotopic studies indicated some *de novo* synthesis as well [47]. *Fainaru et al.* [48] have observed the secretion from perfused rat liver of an apoB-containing discoid lipoprotein but in the density range 1.075–1.175. How it relates to LDL is unclear. Finally, *Fidge and Poulis* [2] report studies in the normal rat indicating that *most* (85–94%) of the LDL with S_f 0–5 (density range 1.040–1.063) is secreted directly into the plasma. In other words, in the rat at least, LDL as we have tentatively defined it does not correspond to a kinetically unique lipoprotein fraction (see 'Introduction'). Whether the LDL comes directly from the liver is not known; *Marsh* [49] fails to find *de novo* LDL secretion from perfused rat liver.

It should be noted that in all of these studies the conclusions regarding the fraction of LDL derived from VLDL are based on the assumption that the labeled VLDL fraction can be treated as an entity, i.e. that all molecules have a similar metabolic fate. If the VLDL fraction includes subfractions of molecules that are preferentially converted to LDL and these molecules are less readily labeled, the true contribution from VLDL will be greater than that calculated. However, no gross heterogeneity of this kind has been documented. The concordant results in three different species strongly suggest that direct secretion of LDL (fig. 1, pathway 5) is a normal process although its magnitude varies from species to species and, as in familial hypercholesterolemia, can be markedly exaggerated under some circumstances.

Does all VLDL apoB go on to LDL before leaving the plasma compartment? This corollary question is probably answered differently according to the species under consideration. In the rat, the answer is decidedly negative. In fact, in the rat *most* VLDL apoB does *not* reach LDL. Instead, most VLDL 'remnants' (IDL) are swept up by the liver and only a small fraction is degraded all the way to the level of LDL [36, 50–52], accounting in part perhaps for the extremely low LDL levels in the rat. In normal man, the reported concordance between flux of apoB in VLDL and LDL mentioned above suggested that no VLDL apoB left the plasma compartment prior to conversion to LDL. However, recent data show that this is not always the case. As much as 50% of the VLDL apoB may leave the plasma before being converted to LDL, being lost from VLDL or, more likely, IDL (fig. 1, pathways 11 and/or 12) [44]. Metabolic patterns may be qualitatively different (or at least quantitatively different) in different patients under different circumstances.

These newer findings are important in reshaping our thinking about lipoprotein interrelationships and interconversions. It is no longer tenable to assume that the liver synthesizes and secretes (in addition to HDL) only VLDL particles of a rather well-defined size and composition from which the IDL and LDL found in plasma are then generated intravascularly. It may be a reasonable approximation under some circumstances but it is an inapplicable oversimplification in others. Why some patients, in particular those with familial hypercholesterolemia, directly secrete large quantities of apoB in IDL or LDL rather than in VLDL is not known. The primary genetic defect in these patients is a loss or modification of cell membrane receptors for LDL, as elegantly shown by the work of *Goldstein and Brown* [8]. The receptor defect has been demonstrated in peripheral cells of several types but it is not yet known whether hepatic cells are abnormal in any way. Hepatic cells might have an analogous receptor determined by the same genome and such a receptor could conceivably play a role in regulating hepatic lipoprotein production. One would not expect, however, that such a regulatory function would lead to qualitative changes in lipoprotein biosynthesis secretion. If we are dealing only with a quantitative difference, that is, an absolute increase in rate of synthesis and secretion however effected, then we must accept that the spectrum of lipoprotein particles secreted by the human liver is not fixed. The situation in familial hypercholesterolemia may only be an extreme example and there may be a spectrum of less drastic perturbations that modify hepatic secretion in different, less obvious ways. For example, normal subjects may secrete some LDL *de novo* but at rates too low to detect by ordinary methods.

In the case of VLDL secretion, we already have evidence that the liver can significantly modify the types of particles it secretes. Carbohydrate induction of hypertriglyceridemia increases the flux of triglycerides in VLDL without necessarily increasing the flux of VLDL apoB [53]. In other words, when the availability of triglyceride is increased, the liver simply puts out larger, triglyceride-rich VLDL [54]. Conversely, it is not unreasonable to propose that when apoB production is stimulated and outruns the available triglyceride, the liver secretes smaller, triglyceride-poor VLDL. In the limit, it might secrete *very* small lipoproteins i.e. IDL or even LDL. These speculations are schematized in figure 1 by indicating the potential of the liver to secrete, in varying proportions, via any of the several pathways 1–5. If apoB production were normal but triglyceride availability were limited, again there might be *de novo* secretion of smaller VLDL, IDL or even LDL. It would be of interest to know whether inhibition of free fatty acid mobilization might invoke or enhance direct hepatic secretion of IDL or LDL.

Once it is recognized that the pattern of primary hepatic secretion products is not necessarily fixed, it becomes possible to explore new hypotheses regarding hyperlipoproteinemia. Mention has already been made of carbohydrate-induced lipemia, where the VLDL level increases without an increase in LDL levels. If, as appears to be the case [53], only VLDL *triglyceride* secretion is increased and *not* VLDL apoB secretion, then no increase in LDL turnover or level should be expected. Thus, it is not necessary to 'explain' the failure of LDL levels to rise, for example, by postulating that the LDL removal mechanisms can respond by increasing removal rate. As a matter of fact, if the rate of VLDL apoB secretion really were increased it would be somewhat surprising to find *no* increase in LDL levels; that would imply an increase in fractional catabolic rate at the *same* LDL level. A point should be stressed here, however, with respect to what we recognize as an elevated level. The range of LDL levels in the normal population is broad. A given patient may have a clearly abnormal VLDL level but an LDL level within what we take as the normal range. Nevertheless, that LDL level may well be distinctly higher than it would have been *in that particular patient* if he had not been overproducing VLDL apoB.

Now that we must include these new pathways in our thinking, the number of possible mechanisms underlying a given lipoprotein phenotype has multiplied enormously. A stripped-down model is shown in figure 2 in which we are considering only the apoB pools in VLDL and LDL. Hyperbetalipoproteinemia could arise in several ways: (1) increased direct LDL secretion (r_4), as in familial hypercholesterolemia; (2) decreased fractional LDL removal rate (r_3), also seen in familial hypercholesterolemia; (3) increased VLDL production (r_1) with normal VLDL-LDL conversion (r_2); (4) normal VLDL production but an increase in fractional conversion to LDL (r_2) compared to that removed by other pathways (r_5), and (5) any appropriate combinations of the above. Phenotypes could include II-A, II-B and IV according to the metabolic error or errors involved and a given phenotype might result from several quite different underlying mechanisms.

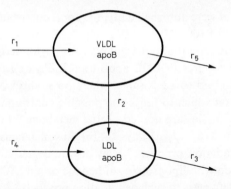

Fig. 2. Scheme indicating the complexity introduced by the double origin of LDL apoB – from VLDL (r_2) or from direct secretion (r_4); and by the double fate of VLDL apoB – going to LDL (r_2) or to tissues (r_5). As discussed in the text, consideration of the intestinal origin of VLDL or LDL and consideration of VLDL triglyceride secretion as a process that can be independently regulated considerably complicates the issue of underlying mechanisms responsible for a given lipoprotein phenotype.

If to the scheme of figure 2 we add the potential for independent control over output of VLDL apoB and of VLDL triglycerides, the number of possible permutations and combinations increases considerably. If we add the potential for intestinal contributions under independent control, the possibilities multiply still further. A given phenotype may arise from several different derangements or combinations of derangements. Consequently, even though a number of kindreds may share a given phenotype and the data in each kindred are compatible with monogenic inheritance, the underlying gene defect need not be the same in all the kindreds [55]. There is nothing more profound here than the recognition that hyperlipoproteinemias may, like other inhertied metabolic diseases, be genetically heterogeneous. The emerging results on the complexities of lipoprotein interrelationships simply make the point more evident. They also indicate the larger number of points at which genetic or environmental influences can modify the expression of a single primary gene defect.

Does all of the LDL apoB in plasma arise from hepatic biosynthesis (whether via VLDL or by direct secretion of LDL itself)? Not necessarily. Some quantity of apoB is delivered into the plasma daily in chylomicrons and VLDL of *intestinal* origin. To what extent this contributes in man to the apoB we find in the LDL fraction is not known. The protein content of human chylomicrons is only about 0.5–1.0% of the total mass and apoB constitutes less than 20%

of that protein [56]. If 100 g of fat were absorbed daily, the apoB contributed to the plasma would be at the most 100–200 mg/day and possibly much less. Turnover of apoB in the LDL fraction is about 1 g/day. Thus chylomicrons would not appear to be a major source of LDL apoB even if chylomicron apoB were quantitatively transferred to LDL (an unlikely possibility). We simply do not have good data on which to judge the possible contribution from intestinal VLDL. Further studies are needed on the metabolic fate of intestinal VLDL and chylomicrons (fig. 1, pathways 14–16). In which density fractions do we find their degradation products? How do the degradation products compare in composition with lipoproteins of hepatic origin? What is the turnover and the steady state mass of such degradation products?

Turnover of LDL

LDL, like other lipoproteins, is a complex of multiple noncovalently linked lipid and protein components. *A priori,* there is no reason to presume that all components share the same metabolic fate nor even that they all leave the plasma compartment as a unit. We are concerned here only with net changes, not exchange reactions by which labeled components (free cholesterol or phospholipids, for example) can equilibrate with identical components in other lipoproteins or tissue stores. The rapid rate at which free cholesterol exchanges among different lipoprotein fractions and between lipoproteins and tissues, for example, all but precludes its use as a marker for LDL turnover and the same is probably true for phospholipids. ApoB, on the other hand, exchanges very slowly if at all and most turnover studies have relied on labeling of the protein moiety. In peripheral cells, it appears that the intact LDL is internalized as a unit by endocytosis, i.e. there does not appear to be any degradation at the capillary endothelium, such as occurs with VLDL, or at the cell surface. To the extent that this mechanism is operative at *all* sites of LDL uptake, the rate of uptake of apoB will reflect the uptake of all the other components. However, the possibility of net transport of lipid components independent of the protein has not been ruled out.

Extravascular LDL Pools

After intravenous injection of labeled LDL, there is an initial rapid phase of disappearance from the plasma followed by a continuing slower phase that closely approximates a true first-order process [57–64]. The initial rapid phase is what would be expected from equilibration with extravascular pools of LDL. In pigs, for example, the early phase has a half-life of about 3 h [60] and the size of the extravascular pool, calculated from analysis of the LDL decay curves, is 20–30% that of the intravascular pool. Analysis of tissues in animals sacrificed at time intervals after injection of ^{125}I-LDL showed that by far the largest part of the inferred extravascular pool resided in the liver [^{125}I-LDL in trapped plasma was corrected for by injecting ^{14}C-dextran just prior to sacrifice, 60]. At all time intervals from 3 to 122 h, the protein ^{125}I in the liver represented a reasonably constant fraction of that in the plasma (15.8 ± SEM 1.9), as would be expected if equilibration is fast relative to the rate of irreversible disappearance from the plasma. In animals receiving ^{125}I-LDL after hepatectomy the disappearance curve was monoexponential instead of biexponential [17], adding further evidence that the liver is the site of the bulk of the extravascular pool of LDL. Studies in man and in other animals support this interpretation of the LDL disappearance curve. The calculated fraction of the total LDL pool accounted for by the extravascular pool is comparable in man, dog and pig — about one-fifth to one-third.

Direct measurement of LDL concentrations in human lymph has yielded values about 10% those in plasma [65, 66]. If all of the extravascular, extracellular fluid space uniformly contained LDL at this concentration, the extravascular LDL pool would indeed represent about 25% of the total. However, as discussed above, the liver appears to account for a disproportionately large fraction of the extravascular pool [60]. The pools referred to here as 'extravascular' are pools that do not mix immediately with LDL in the general circulation. Some of this 'extravascular' pool could reside on capillary walls in the periphery or on sinusoid membranes in the liver rather than within cells. It is possible that the mean LDL concentration in extravascular spaces in the periphery is actually lower than that found in lymph samples. According to current concepts, fluids leave the arteriolar end of the capillary. A large fraction of this returns to the vascular system at the venular end of the capillary and only a small fraction is drained away via the lymphatics. There is still uncertainty about the factors governing movement of macromolecules out of and back into blood capillaries and still more uncertainty with respect to movement into and out of lymphatic capillaries. Thus it seems possible that, because of selectivity at the capillary membranes or varying balances between net fluid transfer and diffusional transfer, concentrations in the lymphatic vessels accessible for sampling may not accurately reflect concentrations bathing the cells in the immediate environment of the blood capillary. Thus lymph samples may be more or less concentrated in lipoproteins than the fluid bathing the majority of cell surfaces. This could be of importance with respect to the regulation of LDL uptake and of endogenous cholesterol biosynthesis in peripheral tissues, as discussed below.

Quantitative Aspects of LDL Turnover

Mathematical techniques for quantification of kinetic studies of lipoprotein turnover are discussed in detail elsewhere in this volume (p. 67). Our discussion will therefore be limited to certain aspects of LDL turnover as they relate to physiologic and pathologic mechanism of apoprotein and cholesterol transport.

In normal man, the net turnover of LDL apoprotein B is generally in the range of 10–20 mg/kg/day. If we assume for the moment that individual LDL particles are removed as units, as appears to be the case for uptake by cells in culture, this corresponds to a net transport of 20–40 mg of LDL cholesterol/ kg/day. In the following section, we discuss the evidence that a significant fraction of LDL transport is attributable to uptake by extrahepatic tissues. Even if only 50% is removed in extrahepatic tissues, 10–20 mg of cholesterol/ kg/day would be delivered to cells that have no capacity to metabolize it – 0.7–1.4 g/day in a 70-kg man. (The adrenal cortex and the gonads can convert cholesterol to more water-soluble products that can be readily retransported and excreted but these pathways account for only a minor fraction of the daily cholesterol turnover in LDL). When the organism is growing, some of this burden of LDL cholesterol can be utilized to synthesize new membranes. It may be that it is in the growing animal that delivery of LDL cholesterol is most important. In the adult, any cholesterol deposited in peripheral tissues must somehow be retransported and ultimately reach the liver – 'reverse cholesterol transport' [67]. Replacement of dying cells does not quantitatively alter the transport problem because the cholesterol content of the cells being replaced must itself somehow also be transported back to the liver.

In addition to the burden of whatever cholesterol is delivered by LDL, virtually all tissues also can synthesize cholesterol. Potentially, then, there is an additional amount of cholesterol to be carried away. The magnitude of extrahepatic synthesis of cholesterol is not known with certainty. Presumably, it depends on the extent to which LDL suppresses endogenous cholesterol synthesis *in vivo* [8]. The concentrations of LDL in lymph as reported by *Reichl et al.* [64, 65] are at or above those needed to suppress cholesterol synthesis in cultured fibroblasts, suggesting that synthesis in peripheral tissues is normally suppressed *in vivo*. Recent studies provide evidence that this is so in the rat [68, 69]. Thus, it seems unlikely that cholesterol synthesized in extrahepatic tissues constitutes a quantitatively important burden in the normal organism.

In patients with familial hypercholesterolemia, the net flux of LDL protein through the plasma compartment is increased [61–64]. In patients with the homozygous form of the disease, the flux is about twice normal and all of the increase is attributable to direct secretion of LDL into the plasma rather than overproduction of VLDL [43]. The basis for this overproduction of LDL is discussed below. Here we only wish to call attention to the fact that the 'burden' of LDL-delivered cholesterol on *some* tissues, possibly the arteries as well, is *increased* despite the fact that many cells in these patients lack the high-affinity receptor mechanism for LDL uptake [8].

Reverse Cholesterol Transport

The mechanism of reverse cholesterol transport *in vivo* remains to be established. Indeed there has thus far been no conclusive demonstration of the process *in vivo*. Most attention has focused on the possibility that HDL (working in concert with lecithin-cholesterol acyltransferase) serves the carrier function, as first proposed by *Glomset* [70]. The evidence supporting this mechanism, attractive in principle, is discussed elsewhere in this volume (p. 41). However, there has been as yet no clear demonstration that it operates *in vivo*. Here we shall limit ourselves to a brief mention of alternative possibilities and some of the technical problems to be faced in assessing them.

Transport in a Fraction Other Than One of the Conventionally Isolated Plasma Lipoprotein Fractions

At a steady state, the net concentration of the fraction of cholesterol with a density greater than 1.21 in human plasma is extremely low (less than 6 mg/dl) but there *is* some there [71, 72]. Serum albumin can bind some cholesterol (as it can almost everything else) and some of the minor components in the fraction with a density greater than 1.21 may also have some cholesterol-binding capacity. We know that cultured cells previously exposed to LDL at high concentrations (and therefore showing an increased cholesterol content) can 'unload' cholesterol to a lipoprotein-depleted serum medium [67, 73–76]. Thus some component or components in the plasma fraction of a density greater than 1.21 has at least some potential capacity to act as acceptor. *Stein et al.* [76] point out that lipoprotein apoproteins, which have a high affinity for lipids, may be present in the lipoprotein-free fraction at sufficient levels to exercise a cholesterol carrier function (in association with phospholipids). Since the concentration of cholesterol in the lipoprotein-free fraction is so low, the flux of cholesterol in it (i.e. the fractional turnover) would have to be quite high if it were to play an important role in reverse cholesterol transport.

Transport within a Subfraction (or Subfractions) of the Conventionally Isolated Plasma Lipoprotein Fractions

HDL. *Glomset's* [70] hypothesis has already been referred to above. HDL as conventionally isolated from plasma (density range 1.063–1.21) or HDL apoproteins in artificial combination with phospholipids increased release of labeled cholesterol from cultured cells [77, 78]. More important, the *net* release of cholesterol from cells previously loaded with cholesterol is increased [76] and the build-up of cholesterol content in cells exposed to LDL is slowed by the presence of HDL in the medium [79, 80]. These *in vitro* results are consonant with *Glomset's* hypothesis but do not establish it. The results may actually underestimate the potential of HDL as acceptor because they utilize steady-state plasma HDL, in which most of the cholesterol is esterified. Nascent HDL particles (i.e. HDL newly secreted by the liver) contain predominantly unesterified cholesterol, as shown by *Hamilton* [81]. These nascent particles are presumably an ideal substrate for lecithin:cholesterol acyltransferase and might therefore represent a subfraction of HDL of special importance in reverse cholesterol transport.

A role for HDL in reverse cholesterol transport has been supported by the finding of cellular accumulation of cholesterol (as esters primarily) in patients with Tangier disease [82]. Recently, *Schaefer and Brewer* [83] have presented an alternative explanation for the cholesterol accumulation. They suggest that chylomicrons in Tangier patients are removed in an abnormal fashion and enter reticuloendothelial cells at a high rate, bringing with them the burden of cholesterol esters and other lipids that accounts for the deposits found.

Finally, mention should be made of the heterogeneity of HDL, particularly the presence of some fraction of molecules containing apoE. Animals fed a cholesterol-rich diet accumulate an abnormal, lipid-rich form of HDL (HDL$_c$) that is rich in apoE [6]. Some apoE-containing molecules are present even in the normal plasma HDL$_1$ fraction [84]. From the work of *Bersot et al.* [12] and *Mahley and Innerarity* [84] we know that lipoproteins containing apoE have a high affinity for the specific LDL receptor on peripheral cells. If acceptor-cell membrane contact is needed for transfer of cholesterol out of the cell, this subfraction of HDL molecules might be of special functional importance. Competitive inhibition of LDL uptake by HDL is discussed below ('Role of LDL in Atherogenesis').

LDL. Ordinarily, we think of LDL as a source of tissue cholesterol. The work of *Goldstein and Brown* [8] suggests that LDL represents a reservoir of cholesterol on which tissues can draw when they require it to replenish cell membranes. Hence, the notion that LDL might function in reverse cholesterol transport seems paradoxical, if not frankly perverse. However, if we keep in mind the possibilities for heterogeneity within density classes as discussed in the Introduction we have to keep our mind open. Indeed, *Sniderman et al.* [85] have presented evidence that some lipoprotein(s) in the fraction with a density between 1.019 and 1.063 deliver cholesterol to the liver in man. Arterial-hepatic vein differences across the liver were positive with respect to LDL cholesterol. The differences were small but statistically significant. In the case of HDL and VLDL, the differences were negative. Corroboration and extension of these interesting results are needed but they are certainly provocative. The possibility that circulating LDL is taken up by and modified by the liver has been previously discussed as a factor in the more rapid catabolism of LDL in hepatectomized animals [17]. It was suggested that in the intact animal, LDL molecules are somehow modified in the liver in a manner that prolongs their lifetime in the circulation [the 'LDL repair hypothesis'; 18]. The large extravascular pool of LDL in the liver could represent LDL undergoing such modification.

IDL. Kinetic studies of the fraction with a density between 1.006 and 1.019 show that it has a longer half-life than that of VLDL. In some species, it is rapidly removed by the liver and therefore seems an unlikely candidate for a role in reverse cholesterol transport. Yet the same arguments presented above with respect to LDL need to be considered here.

VLDL or Chylomicrons. Because of their size, it is unlikely that these large lipoproteins are present at appreciable concentrations outside the vascular bed. They are unlikely to play a role in reverse cholesterol transport from cells other than those in direct contact with plasma (endothelial cells and cells supplied by capillaries containing large fenestrae).

Transport via the Lymphatic System

Reverse cholesterol transport need not occur by way of the plasma compartment. The lymphatic system provides an alternative route. Cells could shed cholesterol onto an acceptor or acceptors in the extracellular fluid and the acceptor(s) might reenter the vascular bed by way of the thoracic duct. Some of the cholesterol in the thoracic duct lymph represents dietary and biliary cholesterol, cholesterol synthesized by the intestine and cholesterol taken up by the intestine from the plasma. How much has its origin in peripheral tissues is not known.

Sites of LDL Degradation

Only the liver can catabolize and excrete cholesterol to a significant extent (the adrenal cortex and gonads account for only a small fraction of the cholesterol turned over daily in LDL). Thus it has been generally assumed – and probably correctly – that LDL cholesterol must ultimately find its way back to the liver. The further assumption – not necessarily correct – has been that *all* of the LDL components return to the liver along with the LDL cholesterol, i.e. that the liver is the 'graveyard' of the LDL. Yet there is little direct evidence in the literature to support this assumption.

Entenman et al. and *Zilversmit and Bollman* [87] carried out pioneering studies of lipoprotein metabolism in which they injected ^{32}P-phosphate and followed the specific radioactivity of plasma phospholipids as a function of time. They found that hepatectomy in rats and in dogs drastically reduced the incorporation of ^{32}P into lipoproteins, in keeping with the now generally accepted primary role of the liver in lipoprotein synthesis and secretion. They also found that when the liver was removed *after* labeling the animal with ^{32}P, the specific radioactivity of the plasma phospholipids fell very slowly, again indicating that extrahepatic tissues make at most a minor contribution to plasma phospholipids. *Hotta and Chaikoff* [88] labeled the plasma cholesterol pool of rats and showed that functional hepatectomy almost completely arrested the previously log-linear fall in plasma cholesterol specific activity. Again, this shows that the liver is the primary source of plasma cholesterol, i.e. no newly synthesized labeled cholesterol enters the plasma after hepatectomy. However, no data on *net* plasma cholesterol levels were reported and no conclusions regarding removal can be drawn.

Entenman et al. [86] measured both specific activity of phospholipids and net plasma concentrations before and after evisceration in dogs (no portal vein or hepatic artery blood supply to the liver). The rate of fall of ^{32}P in phospholipids was sharply reduced or arrested. Total plasma phospholipids decreased very little or not at all and it was concluded that the liver was both the source and the sink for plasma phospholipids. The first point to emphasize is that 80–90% of the cholesterol and phospholipid in normal dog plasma is in the HDL fraction. Lipoproteins were of course not subfractionated in these early studies and the results must reflect primarily the metabolism of HDL – not LDL (the same is true for the rat studies). Second, the longest time interval of sampling after hepatectomy or evisceration was 10 h. If the half-life of HDL in the dog is about 2 days, levels would only be expected to fall by 14% in 10 h even if removal rates were not affected by hepatectomy. Thus these studies, while carefully done and cautiously interpreted, do not rule out significant extrahepatic degradation of LDL or, for that matter, of HDL. They do show that biosynthesis and secretion of lipoproteins by extrahepatic tissues were quantitatively minor under the conditions used.

Hay et al. [19] studied the metabolism of ^{125}I-LDL in perfused rat livers. They found that the liver removed ^{125}I-LDL from the perfusate but it was not clear what fraction of the ^{125}I-LDL taken up was in fact degraded. As discussed above, the liver contains a large pool of LDL that can equilibrate rather rapidly with plasma LDL [60]. Thus, the initial disappearance of ^{125}I-LDL from the perfusate could reflect equilibration rather than degradation. While the quantitative aspects of this study are incomplete, the finding of some increase in unbound ^{125}I in the perfusate suggests that the liver has at least some capacity to degrade LDL. This is also indicated by studies of cultured hepatocytes. Studies by *Pangburn et al.* [21] show that rat and pig hepatocytes can degrade ^{125}I-LDL. *Breslow et al.* [20] also observed LDL degradation in rat hepatocyte cultures although it was less than that in rat fibroblast cultures.

In an attempt to directly assess the role of the liver in LDL degradation *in vivo, Sniderman et al.* [17] and *Steinberg et al.* [18] determined the fractional catabolic rate of ^{125}I-LDL in pigs and in dogs before and immediately after total hepatectomy, expecting to find that disappearance of ^{125}I-LDL would be arrested or drastically slowed in the absence of the liver. Instead they found that the rate of disappearance of ^{125}I-LDL was *not* reduced by hepatectomy in either species. Further, the *net* plasma levels of LDL decreased at the same rate as the decrease in ^{125}I-LDL, i.e. the specific radioactivity did not change. The latter finding provides additional evidence that the liver is the major site of origin of LDL. The continuing disappearance of both LDL protein and LDL radioactivity showed that extrahepatic tissues have the capacity to degrade LDL at a rate comparable to that seen in the intact animal. This finding, as pointed out by the authors, 'does not necessarily rule out some degree of LDL catabolism by the normal liver'. Quite possibly, metabolic changes in the hepatectomized animals may in some way enhance the rate of

LDL catabolism by extrahepatic tissues. This possibility is strengthened by the fact that after hepatectomy the observed fractional catabolic rate for LDL was actually somewhat greater than that in the intact animals. Thus, the liver could, in the intact animal, contribute to LDL degradation and yet hepatectomy need not reduce the overall rate of LDL degradation. An alternative way of interpreting the result is to propose that in the intact animal the liver somehow acts on LDL so as to prolong its lifetime in the circulation ['LDL repair hypothesis'; 18]. These considerations are highly speculative and further studies are needed to resolve the role of the liver in LDL metabolism. Here we are concerned only with the major conclusion drawn – that extrahepatic tissues probably play a major role in LDL catabolism.

If LDL is degraded peripherally, it must follow that some or many peripheral cell types have a relatively high capacity for degrading LDL. This corollary hypothesis is supported by studies in several cell lines. *Weinstein et al.* [67] incubated smooth muscle cells derived from swine aorta with ^{125}I-LDL at a concentration 10% of that in swine plasma – a concentration approximating that presumed to be present in the extracellular, extravascular space [65, 66]. The observed degradation rate per milligram of cell protein was actually 5–10 times the *in vivo* degradation rate in intact swine [60] expressed in the same terms. Comparison of rates of LDL degradation by human cell lines (fibroblasts, smooth muscle cells, endothelial cells) with rates of LDL degradation in intact man lead to a similar conclusion [89]. *Brown and Goldstein* [90] have shown that the rate of LDL degradation in cultured cells varies depending on whether the LDL receptors are fully expressed or not. However, even when the LDL receptor number is at a minimum, as when the cells are incubated in the presence of LDL or free cholesterol in the medium, the rate of LDL degradation is still high enough to be comparable to the *in vivo* rate. LDL degradation has been studied in the isolated perfused swine heart [*T.E. Carew*, unpubl.]. In this system, the extravascular LDL concentration is determined by the permeability of the capillary bed and should approximate the physiological value. Rates of LDL degradation were 30–50% of the rate in intact swine (per unit mass of tissue). Thus studies in isolated tissues are compatible with the proposal that extrahepatic degradation may be appreciable.

Sites of apoprotein degradation in the intact animal are difficult to determine because the degradation products, mostly free amino acids, very quickly leave the tissue and are excreted. Consequently, the amounts of tracer found in the tissue need not be an index of the degradative activity of that tissue. *Stein et al.* [91] exploited the ability of chloroquine to inhibit lysosomal degradation and thus obtained a measure of initial rates of lipoprotein degradation by the liver. Rats were pretreated with chloroquine, given ^{125}I-labeled human LDL, and sacrificed after 50% of the ^{125}I-LDL had disappeared from the plasma. In control animals only 4% and in chloroquine-treated animals only 10% of the injected dose was recovered in the liver. In contrast, after injection of ^{125}I-labeled human VLDL as much as 80% of the dose was recovered in the liver of chloroquine-treated rats at 45 min. Taken together, the findings summarized above support the conclusions that: (1) the liver is not the *exclusive* site of LDL degradation, although it may contribute; (2) extrahepatic tissues play a *prominent* role in LDL degradation, although the relative contributions of different tissues remains to be established.

Recently, *Pittman and Steinberg* [92] have proposed a novel approach to determining the relative contributions of various tissues to overall LDL degradation in the intact animal. The method takes advantage of the fact that certain classes of compounds accumulate in lysosomes. For example, sucrose is taken up by pinocytosis by cells in culture and retained in the lysosomes, leaking only very slowly [93, 94]. *Pittman and Steinberg* covalently coupled ^{14}C-sucrose to ^{125}I-LDL and incubated the doubly labeled lipoprotein with cultured human skin fibroblasts. Degradation was measured in the usual way by determining the amount of ^{125}I appearing in the medium in TCA-soluble form. The amount of ^{14}C accumulated intracellularly in TCA-soluble form was taken as an independent measure of LDL degradation, assuming that after lysosomal degradation of the LDL protein the conjugated sucrose moiety would accumulate quantitatively in the lysosomes. The agreement between the two measures of degradation was excellent.

If the *in vitro* results apply *in vivo*, each tissue will accumulate progressively an amount of TCA-soluble ^{14}C-sucrose (or alternative lysosomotropic marker) proportional to the number of lipoprotein molecules it has degraded up to the time of sacrifice. In other words, the build-up of ^{14}C-sucrose provides a cumulative tally, each molecule degraded leaving behind its 'calling card' – ^{14}C-sucrose. An absolute measure of the amount of protein degraded would require relating the tissue ^{14}C to the integrated specific radioactivity of the serum protein studied over the duration of the experiment. However, the *relative* contributions of the different tissues should be simply proportional to their ^{14}C content whatever the shape of the plasma-specific activity curve. Studies in swine show that about 40% of LDL degradation occurs in liver and 60% extrahepatically. The adrenal, liver and adipose tissue were the most active sites of degradation expressed per gram wet weight (*Pittman, Attie and Steinberg*, unpublished).

Factors Determining Steady-State Plasma Levels of LDL

Plasma LDL levels are determined by the balance between rates of production and rates of removal. We have already discussed ('Origin of LDL') the potential dual origin of LDL – from VLDL and by *de novo* secretion – and the potential variability in the fraction of VLDL apoB converted to LDL as opposed to that leaving the plasma by other routes (fig. 2). Intensive research on the relative importance of these alternative pathways in patients with hyperlipoproteinemia, primary and secondary, can be expected in the coming years but little more can be said at this time. In this section we will confine ourselves to a discussion of a few examples of genetically determined abnormal LDL metabolism and how they fit into the context of current concepts as presented above.

In *abetalipoproteinemia,* there is an essentially complete absence of chylomicrons, VLDL and LDL – all of the lipoprotein fractions containing apoB [95]. The inborn error probably represents a failure to synthesize a functional apoB or failure to conjugate it with lipids. Accelerated removal does not appear to be the mechanism as shown by the results of *Lees and Ahrens* [96]. They transfused concentrated normal LDL into an abetalipoproteinemic patient to increase LDL levels and then measured the rate of fall when LDL infusions were discontinued. The rate of fall was only slightly greater than the rate of decay of ^{125}I-LDL in normal subjects. In familial *hypobetalipoproteinemia,* the metabolic error once again appears to be related to a decrease in production. The fractional catabolic rate of LDL is normal or only moderately increased but total flux is markedly decreased [97–99]. The modest but definite increase in fractional catabolic rate for LDL in these patients with very low LDL levels could reflect a maximal induction of cell membrane LDL receptors, by the mechanism described by *Brown and Goldstein* [96]. They showed that in fibroblasts exposed to high levels of LDL (or free cholesterol), the number of expressed LDL receptors is minimized while in cells incubated in lipoprotein-free medium the number of receptors is maximized. Thus the cells in patients with abetalipoproteinemia or hypobetalipoproteinemia may have an increased number of high-affinity receptors and an increased ability to catabolize LDL at low concentrations. However, their extremely low LDL levels must be attributed primarily to a deficiency in secretion of LDL and/or its precursor, VLDL.

At the opposite end of the spectrum, we have *familial hypercholesterolemia,* in which the primary defect lies in LDL catabolism. Both in heterozygotes [59, 64] and in homozygotes [61, 62], the fractional catabolic rate for

LDL is reduced. While it is true that LDL production is increased, particularly in the homozygotes [43] but perhaps to some extent also in heterozygotes [64], it seems more likely that this is secondary or else an accompanying consequence of the primary receptor defect described by *Goldstein and Brown* [100]. Details of their careful and comprehensive elucidation of the receptor defect are discussed in a recent review [8] and elsewhere in this volume (p. 216). Suffice it to say that the powerful genetic evidence from their studies of fibroblasts from heterozygotes and homozygotes, and particularly their elegant characterization of a new allelic mutation affecting the *internalization* function of the receptor [101], make a convincing case for the primacy of the receptor defect.

How then do we account for the overproduction of LDL? The studies of *Soutar et al.* [43] show that the increment in LDL turnover above the normal is largely accounted for by *de novo* LDL secretion rather than VLDL secretion. Above we alluded to the possibility that the nature of the apoB-containing lipoproteins secreted by the liver may vary according to the relative availability of triglycerides and cholesterol ('Origin of LDL'). If the extrahepatic tissues of patients with familial hypercholesterolemia overproduce cholesterol (non-repressed cells) and that newly synthesized cholesterol is being returned to the liver, it might be a trigger for increased lipoprotein production, specifically the production of cholesterol-rich lipoproteins, i.e. LDL. An alternative view is that the liver cell has a receptor determined by the same defective gene that determines production of the extrahepatic cell receptor and that in the liver this receptor somehow regulates the rates of LDL production. *Fogelman et al.* [102] propose that the extrahepatic cells in patients with familial hypercholesterolemia 'leak' cholesterol more readily and that this accounts, at least in part, for the derepression of HMG CoA reductase. It might, as discussed above, also be a stimulus to hepatic LDL production. Unfortunately, we know too little about the factors regulating the rates and patterns of lipoprotein production by the liver (and by the intestine), particularly the possible feedback controls on these processes.

Role of LDL in Atherogenesis

A wealth of experimental, epidemiological and clinical evidence links the LDL fraction of plasma to the pathogenesis of atherosclerosis. However, it is not at all clear what the mechanisms are by which LDL accelerates the atherogenic process. Proponents of the 'insudation theory' have looked upon this relationship between LDL and atherogenesis from the point of view that the arterial wall is a tissue exposed to and having to deal with various rates of LDL infiltration. The LDL fraction delivers cholesterol and cholesterol esters to the artery wall at a rate determined by the concentration of LDL in the plasma and also as a function of factors determining the rate of infiltration (blood pressure, endothelial permeability and other factors). If the rate of cholesterol delivery exceeds the ability of the tissue to metabolize and remove it, cholesterol accumulation follows and somehow favors atherogenesis. This simplistic view must be reduced to the level of cellular mechanisms and, ultimately, molecular mechanisms before it becomes sufficiently well-defined to allow for critical evaluation. Since cell proliferation, probably proliferation of smooth muscle cells [103], is a characteristic feature of atheroma formation, a complete hypothesis must account for this also, i.e. one must invoke mechanisms that account for cell proliferation. *Fischer-Dzoga and Wissler* [104] have proposed that LDL (at least LDL from hyperlipoproteinemic animals or hyperlipoproteinemic patients) can itself stimulate the proliferation of smooth muscle cells. We do not have evidence that this occurs *in vivo* but if so hyperbetalipoproteinemia might be an adequate basis for both the proliferative aspects of the lesion and the characteristic accumulation of lipid in the lesions. Another possibility is that hyperbetalipoproteinemia leads both to proliferation and to lipid accumulation but by different mechanisms. *Ross and Harker* [105] have shown that the survival time of platelets is reduced in hyperlipoproteinemic animals. They propose that an increased level of LDL favors denudation of the endothelial lining. This could lead to platelet aggregation, release of the platelet growth factor of *Ross et al.* [106] and thus to proliferation of the smooth muscle cells. Concurrently, exposure of these cells to much higher concentrations of LDL (because of the loss of the endothelial barrier) could account for the accumulation of cholesterol in the cells because the 'burden' is in excess of the capacity of the smooth muscle cells to rid themselves of cholesterol taken in with LDL.

Assessment of endothelial integrity *in vivo* is difficult. The very fact that it is almost impossible to isolate arteries and study them *in vitro* without damaging the endothelial layer suggests that some degree of endothelial damage may be unavoidable *in vivo* as it is *in vitro*. In that case, all arterial segments are always at some risk because of continuing focal endothelial damage, damage which is ordinarily repaired continuously and therefore may never be extensive. Thus, it is still a possibility that we do not need to invoke *enhanced* endothelial damage in order to account for excessive 'load' of lipoprotein uptake in various segments of the arterial tree. In short, it is still a possibility that atherosclerosis may occur, and proceed at an accelerated rate with hyperbetalipoproteinemia, even without unwonted breaching of endothelial integrity. Of course, if the endothelial lining is severely injured, the events briefly summarized here proceed at a markedly accelerated pace [107].

De Duve [108] has proposed that the capacity of the lysosomes in arterial smooth muscle cells to degrade LDL may be a limiting factor determining whether there will or will not be atherogenesis. He points out that the activity of lysosomal enzymes, in the artery wall of rabbits, a species highly susceptible to atherosclerosis, is much less than that in the arterial wall of other species. The thesis is that atherosclerosis will not progress if the cells can keep up with the rate at which LDL is delivered but will progress when there is a disproportion between delivery and the capacity of the cells to degrade.

Why are the smooth muscle cells in atherosclerotic lesions unable to protect themselves from becoming loaded with cholesterol? Like fibroblasts, they have a finite, saturable number of high-affinity LDL receptors and LDL regulates endogenous cholesterol synthesis [67, 109]. If the receptor number is under the same kind of regulation described in fibroblasts, there should be a ceiling on LDL uptake. An increase in cell cholesterol content should lead to down-regulation of the number of LDL receptors and a consequent reduction in rate of LDL uptake. Evidently, the system is not functioning effectively in this way. We suggest that in this situation, when LDL levels are elevated, LDL uptake by way of the *low* affinity mechanism is of great significance [89, 110]. This must be the case in patients with the receptor-negative form of familial hypercholesterolemia [8]. Despite the absence of high-affinity receptors, total daily LDL turnover is actually increased in these patients [43, 61–64]. Fibroblasts from these patients *do* degrade LDL, although at a markedly reduced rate [100, 110, 111]. The rate of degradation is in excess of that expected from fluid endocytosis only, indicating that adsorptive endocytosis from low-affinity binding sites must be involved [110]. In the intact patient, uptake via this pathway can account for at least some of the observed LDL degradation. Cells in some body tissues may not have high-affinity receptors at all or have very few of them, as pointed out by *Brown and Goldstein* [112]. In the face of hyperbetalipoproteinemia, the relative contribution of such tissues to LDL degradation would become proportionately greater. Thus some combination of LDL uptake from low-affinity sites in cells that normally bear receptors and in cells that normally lack receptors may account for the observed LDL degradation in familial hypercholesterolemia.

Another point that should be considered in relation to the atherogenicity of LDL is that LDL molecules may undergo structural modification that affect their uptake and metabolism. A striking example of this phenomenon is the profound change in metabolism of glycoproteins induced by desialylation [16]. As discussed briefly above, this may not be an important factor in LDL metabolism [22] but other molecular changes could be significant. *Moore et al.* [113] have shown that conformational changes in albumin can change its rate of uptake by yolk sac cells more than tenfold. *Carew et al.* [114] report that mild tryptic treatment of LDL (releasing about 20% of the protein but no lipid) yields a soluble lipoprotein that is degraded by cells from patients with homozygous familial hypercholesterolemia at 4–20 times the rate seen with native LDL. In cells from normal subjects, however, the degradation of the trypsin-treated LDL was not altered. *Basu et al.* [115] have shown that 'cationized LDL', i.e. LDL covalently modified so as to markedly increase its positive charge, is rapidly taken up by fibroblasts and that the uptake is receptor-independent. If LDL molecules undergo analogous changes *in vivo* the more rapid (and presumably nonspecific) uptake of such 'damaged' molecules might help account for atherogenesis in familial hypercholesterolemia [114]. Since the half-life of LDL in familial hypercholesterolemic patients is prolonged [61–64], the opportunities for 'damage' are greater in them. Attention should be drawn also to the fact that LDL is present in the intima at high concentration [116]. Some of this LDL is bound tightly to elements of the extracellular matrix (elastin and glycosaminoglycans) and may reside in the extracellular intimal space for an appreciable time. During this time, it may undergo denaturation or enzymatically induced alterations that allow it to be taken up rapidly by cells once released from its binding sites.

Studies on the functioning of the LDL receptor mechanism *in vivo* are still limited in number. That they do function to suppress cholesterol synthesis in extrahepatic tissues is suggested by studies in the rat referred to above [68, 69]. Further work is needed to assess their function in respect to LDL catabolism in various tissues and, specifically, to test whether changes in the efficiency of LDL uptake in the arterial wall contribute to premature atherosclerosis even in patients with normal plasma LDL levels. In this connection, the now well-established 'protective' effect of high levels of HDL against premature clinical atherosclerosis should be mentioned [117–120]. Two hypotheses have been proposed that might explain this at the cellular level: (1) that HDL favors reverse cholesterol transport from extrahepatic tissues as discussed above ('Turnover of LDL; Reverse Cholesterol Transport'); (2) that some subfractions of HDL reduce the rate of LDL uptake by extrahepatic tissues. The latter proposal stems from cell culture studies showing that high levels of HDL reduce the binding, uptake and degradation of LDL in several cell types [79, 80, 121, 122]. This interaction may reflect primarily the effects of HDL molecules containing apoE, shown to have a high affinity for the LDL receptor [12–14], but other effects may be involved since interaction was seen also in fibroblasts from patients with familial hypercholesterolemia [110]. Additional studies are needed to explore other possible interactions as well as metabolic and hormonal factors that may influence LDL uptake, receptor-mediated or receptor-independent.

One final caveat should be offered. We are still uncertain about the relative importance of cellular uptake of LDL as an element in the pathogenesis of atherosclerosis. Some of the cholesterol ester in early lesions and most of it in late lesions is found extracellularly and has a fatty acid composition like that of plasma LDL, implying that it has not been metabolized by the cells [116]; intracellular cholesterol esters are relatively rich in oleic acid, the result of lysosomal hydrolysis followed by reesterification [8]. Quite possibly, uptake by smooth muscle cells is not a crucial element in atherogenesis but accumulation extracellularly is. As turns out to be the case with most scientific dichotomies, probably both are. While we know a great deal about the mechanisms of LDL formation, turnover and cellular catabolism, we are still not certain how, at the cellular and molecular level, hyperbetalipoproteinemia is atherogenic!

Acknowledgements

The author acknowledges support for him self and his colleagues from the Specialized Center of Research on Arteriosclerosis, HL-14197, and HL-07276 (both awarded by the National Heart, Lung and Blood Institute) as well as grant GM-17702 (awarded by the National Institute of General Medical Sciences and Stroke).

References

1 Goebel, R.; Garnick, M., and Berman, M.: A new model for low density apolipoprotein kinetics: evidence for two labeled moieties. Circulation 54: suppl. II, p. 4 (1976).
2 Fidge, N. H. and Poulis, P.: Metabolic heterogeneity in the formation of low density lipoprotein from very low density lipoprotein in the rat: evidence for the independent production of a low density lipoprotein subfraction. J. Lipid Res. 19: 324–349 (1978).
3 Lee, D. M. and Alaupovic, P.: Physicochemical properties of low-density lipoproteins of normal human plasma: evidence for the occurrence of lipoprotein B in associated and free forms. Biochem. J. 137: 155–167 (1974).

4 Eisenberg, S.: Lipoprotein metabolism and hyperlipemia; in Paoletti and Gotto, jr., Atherosclerosis reviews, vol. 1, pp. 23–60 (Raven Press, New York 1976).
5 Deckelbaum, R.; Eisenberg, S.; Barenholz, Y., and Olivercrona, T.: Production of low density lipoprotein-like ('LDL') particles *in vitro*. Circulation 57: suppl. III, p. 56 (1977).
6 Mahley, R.W.; Weisgraber, K.H.; Innerarity, T.; Brewer, H.B., jr., and Assman, G.: Swine lipoproteins and atherosclerosis. Changes in the plasma lipoproteins and apoproteins induced by cholesterol feeding. Biochemistry, N.Y. 14: 2817–2823 (1975).
7 Edgington, T.S.; Chisari, F.V., and Curtiss, L.K.: Regulation of lymphocyte function in HBV infection; in Miescher, Immunopathology. VIIth Int. Symp., pp. 173–191 (Schwabe, Basel 1976).
8 Goldstein, J.L. and Brown, M.S.: The low density lipoprotein pathway and its relation to atherosclerosis. A. Rev. Biochem. 46: 897–930 (1977).
9 Jackson, R.L.; Morrisett, J.D., and Gotto, A.M., jr.: Lipoprotein structure and metabolism. Physiol. Rev. 56: 259–316 (1976).
10 Osborne, J.C., jr. and Brewer, H.B.: The plasma lipoproteins; in Anfinsen, Edsall and Richards, Advances in protein chemistry, pp. 253–337 (Academic Press, New York 1977).
11 Bradley, W.A.; Rohde, M.F.; Jackson, R.L., and Gotto, A.M., jr.: Aggregation states of the cyanogen bromide peptides of human low density lipoproteins. Fed. Proc. Fed. Am. Socs. exp. Biol. 36: 829 (1977).
12 Bersot, T.P.; Mahley, R.W.; Brown, M.S., and Goldstein, J.L.: Interaction of swine lipoproteins with the low density lipoprotein receptor in human fibroblasts. J. biol. Chem. 251: 2395–2398 (1976).
13 Mahley, R.W. and Innerarity, T.L.: Interaction of canine and swine lipoproteins with the low density lipoprotein receptor of fibroblasts as correlated with heparin/manganese precipitability. J. biol. Chem. 252: 3980–3986 (1977).
14 Mahley, R.W. and Innerarity, T.L.: Properties of lipoproteins responsible for high affinity binding to cell surface receptors of fibroblasts and smooth muscle cells. Proc. 6th Int. Symp. on Drugs Affecting Lipid Metabolism; in Kritchevsky, Paoletti and Holmes, Advances in experimental medicine and biology 109: 99–127 (Plenum Press, New York 1978).
15 Swaminathan, N. and Aladjem, F.: The monosaccharide composition and sequence of the carbohydrate moiety of human serum low density lipoproteins. Biochemistry, N.Y. 15: 1516–1522 (1976).
16 Ashwell, G. and Morell, A.G.: The role of surface carbohydrates in the hepatic recognition and transport of circulating glycoproteins. Adv. Enzymol. 41: 99–128 (1974).
17 Sniderman, A.D.; Carew, T.E.; Chandler, J.G., and Steinberg, D.: Paradoxical increase in rate of catabolism of low-density lipoproteins after hepatectomy. Science 183: 526–528 (1974).
18 Steinberg, D.; Carew, T.E.; Chandler, J.G., and Sniderman, A.D.: The role of the liver in metabolism of plasma lipoproteins; in Lundquist and Tygstrup, Regulation of hepatic metabolism, pp. 144–156 (Munksgaard, Copenhagen 1974).
19 Hay, R.V.; Pottenger, L.A.; Reingold, A.L.; Getz, G.S., and Wissler, R.W.: Degradation of I^{125}-labeled serum low density lipoprotein in normal and estrogen-treated male rats. Biochem. biophys. Res. Commun. 44: 1471–1477 (1971).
20 Breslow, J.L.; Spaulding, D.R., and Lothrop, D.A.: Lipoprotein binding, de-

gradation and regulation of sterol production in rat liver and fibroblast cell cultures. Circulation 52: suppl. II, p. 59 (1975).
21 Pangburn, S.; Weinstein, D.B., and Steinberg, D.: Degradation of human low density (LDL) and high density (HDL) lipoproteins by cultured hepatocytes. Fed. Proc. Fed. Am. Socs exp. Biol. 37: 1482 (1978).
22 Attie, A.; Weinstein, D.B.; Freeze, H.; Pittman, R.C., and Steinberg, D.: Unaltered turnover of neuraminidase-treated low density lipoprotein in the pig. Biochem. J. (1979, in press).
23 Steinberg, D.; Nestel, P.J.; Weinstein, D.B.; Remaut-Desmeth, M., and Chang, C.M.: Interactions of native and modified human low density lipoproteins with human skin fibroblasts. Biochim. biophys. Acta 528: 199–212 (1978).
24 Gustafson, A.: New method for partial deplipidization of serum lipoproteins. J. Lipid Res. 6: 512–517 (1965).
25 Fisher, W.R. and Shireman, R.B.: Demonstration of the specificity for binding by the low density lipoprotein fibroblast receptor for apolipoprotein B. Fed. Proc. Fed. Am. Socs exp. Biol. 36: 935 (1977).
26 Higgins, M.J.P.; Lecamwasam, D.S., and Galton, D.J.: A new type of familial hypercholesterolemia. Lancet ii: 737–740 (1975).
27 Myant, N.B.; Reichl, D.; Thompson, G.R.; Higgins, M.J.P., and Galton, D.J.: The metabolism *in vivo* and *in vitro* of plasma low-density lipoprotein from a subject with inherited hypercholesterolemia. Clin. Sci. mol. Med. 51: 463–465 (1976).
28 Malloy, M.S. and Kane, J.P.: Normotriglyceridemic abetalipoproteinemia – Clinical and biochemical features of a new syndrome. Pediat. res. 11: 519 (1977).
29 Gitlin, D.; Cornwall, D.G.; Na'Casato, D.; Oncley, J.L.; Hughes, W.L., and Janeway, C.A.: Studies on the metabolism of plasma proteins in the nephrotic syndrome. II. The lipoproteins. J. clin. Invest. 37: 172–186 (1958).
30 Havel, R.J.: Conversion of plasma free fatty acids into triglycerides of plasma lipoprotein fractions in man. Metabolism 10: 1031–1034 (1961).
31 Havel, R.J.; Felts, J.M., and Van Duyne, C.M.: Formation and fate of endogenous triglycerides in blood plasma of rabbits. J. Lipid Res. 3: 297–308 (1962).
32 Fried, M.; Wilcox, H.G.; Faloona, G.R.; Eoff, S.P.; Hoffman, M.S., and Zimmerman, D.: The biosynthesis of plasma lipoproteins in higher animals. Comp. Biochem. Physiol. 25: 651–661 (1968).
33 Quarfordt, S.H.; Frank, A.; Shames, D.M.; Berman, M., and Steinberg, D.: Very low density lipoprotein triglyceride transport in type IV hyperlipoproteinemia and the effects of carbohydrate-rich diets. J. clin. Invest. 49: 2281–2297 (1970).
34 Bilheimer, D.W.; Eisenberg, S., and Levy, R.I.: The metabolism of very low density lipoprotein proteins. I. Preliminary *in vitro* and *in vivo* observations. Biochim. biophys. Acta 260: 212–221 (1972).
35 Sigurdsson, G.; Nicoll, A., and Lewis, B.: Conversion of very low density lipoprotein to low density lipoprotein: a metabolic study of apolipoprotein B kinetics in human subjects. J. clin. Invest. 56: 1481–1490 (1975).
36 Faergeman, O.; Sata, T.; Kane, J.P., and Havel, R.J.: Metabolism of apoprotein B of plasma very low density lipoproteins in the rat. J. clin. Invest. 56: 1396–1403 (1975).
37 Eisenberg, S.; Bilheimer, D.W., and Levy, R.I.: The metabolism of very low density lipoprotein proteins. II. Studies on the transfer of apoproteins between plasma lipoproteins. Biochim. biophys. Acta 280: 94–104 (1972).

38 Eisenberg, S.; Bilheimer, D.W.; Lindgren, F.T., and Levy, R.I.: On the metabolic conversion of human plasma very low density lipoprotein to low density lipoprotein. Biochim. biophys. Acta *326:* 361–377 (1973).

39 Barter, P.J. and Nestel, P.J.: Precursor-product relationship between pools of very low density lipoprotein triglyceride. J. clin. Invest. *51:* 174–180 (1972).

40 Phair, R.D.; Hammond, M.G.; Bowden, J.A.; Fried, M.; Fisher, W.R., and Berman, M.: A preliminary model for human lipoprotein metabolism in hyperlipoproteinemia. Fed. Proc. Fed. Am. Socs exp. Biol. *34:* 2263 (1975).

41 Eisenberg, S.; Bilheimer, D.W.; Lindgren, F.T., and Levy, R.I.: On the apoprotein composition of human plasma very low density lipoprotein subfractions. Biochim. biophys. Acta *260:* 329–333 (1972).

42 Felts, J.M. and Itakura, H.: The mechanism of assimilation of remnant particles by liver – a new theory. Circulation *52:* suppl. II, p. 38 (1975).

43 Soutar, A.K.; Myant, N.B., and Thompson, G.R.: Simultaneous measurement of apolipoprotein B turnover in very-low- and low-density lipoproteins in familial hypercholesterolaemia. Atherosclerosis *28:* 247–256 (1977).

44 Janus, E.; Wootton, R.; Nicoll, A.; Turner, P., and Lewis, B.: Quantitation of very low density (VLDL) to low density lipoprotein (LDL) in normal and hyperlipidaemic man. Circulation *56:* suppl. III, p. 21 (1977).

45 Illingworth, D.R.: Metabolism of lipoproteins in nonhuman primates. Studies on the origin of low density lipoprotein apoprotein in the plasma of the squirrel monkey. Biochim. biophys. Acta *388:* 38–51 (1975).

46 Nakaya, N.; Chung, B.H., and Taunton, O.D.: Synthesis of plasma lipoproteins by the isolated perfused liver from the fasted and fed pig. J. biol. Chem. *252:* 5258–5261 (1977).

47 Nakaya, N.; Chung, B.H.; Patsch, J.R., and Taunton, O.D.: Synthesis and release of low density lipoproteins by the isolated perfused pig liver. J. biol. Chem. *252:* 7530–7533 (1977).

48 Fainaru, M.; Felker, T.E.; Hamilton, R.L., and Havel, R.J.: Evidence that a separate particle containing B-apoprotein is present in high-density lipoproteins from perfused rat liver. Metabolism *26:* 999–1004 (1977).

49 Marsh, J.B.: Lipoproteins in a nonrecirculating perfusate of rat liver. J. Lipid Res. *15:* 544–550 (1974).

50 Faergeman, O. and Havel, R.J.: Metabolism of cholesteryl esters of rat very low density lipoproteins. J. clin. Invest. *55:* 1210–1218 (1975).

51 Mjøs, O.D.; Faergeman, O.; Hamilton, R.L., and Havel, R.J.: Characterization of remnants of lymph chylomicrons and lymph and plasma very low density lipoproteins in 'supradiaphragmatic' rats. Eur. J. clin. Invest. *4:* 382–383 (1974).

52 Eisenberg, S. and Rachmilewitz, D.: Metabolism of rat plasma very low density lipoprotein. I. Fate in circulation of the whole lipoprotein. Biochim. biophys. Acta *326:* 378–390 (1973).

53 Melish, J.; Le, N.A.; Ginsberg, H.; Brown, W.V., and Steinberg, D.: Effect of high carbohydrate diet on very low density lipoprotein apoprotein-B and triglyceride production. Circulation *56:* suppl. III, p. 5 (1977).

54 Ruderman, N.B.; Jones, A.L.; Krauss, R.M., and Shafrir, E.: A biochemical and morphologic study of very low density lipoproteins in carbohydrate-induced hypertriglyceridemia. J. clin. Invest. *50:* 1355–1368 (1971).

55 Goldstein, J.L.; Schrott, H.G.; Hazzard, W.R.; Bierman, E.L., and Motulsky, A.G.: Hyperlipidemia in coronary heart disease. II. Genetic analysis of lipid levels in 176 families and delineation of a new inherited disorder, combined hyperlipidemia. J.clin. Invest. 52: 1544–1568 (1973).

56 Kostner, G. and Holasek, A.: Characterization and quantiation of the apolipoproteins from human chyle chylomicrons. Biochemistry, N.Y. 11: 1217–1223 (1972).

57 Walton, K.W.; Scott, P.J.; Jones, J.V.; Fletcher, R.F., and Whitehead, T.: Studies on low density lipoprotein turnover in relation to atromid therapy. J.Atheroscler.Res. 3: 396–414 (1963).

58 Scott, P.J. and Hurley, P.J.: Effect of clofibrate on low density lipoprotein turnover in essential hypercholesterolemia. J.Atheroscler.Res. 9: 25–34 (1969).

59 Langer, T.; Strober, W., and Levy, R.I.: The metabolism of low density lipoprotein in familial type II hyperlipoproteinemia. J.clin.Invest. 51: 1528–1536 (1972).

60 Sniderman, A.D.; Carew, T.E., and Steinberg, D.: Turnover and tissue distribution of ^{125}I-labeled low density lipoprotein in swine and dogs. J.Lipid Res. 16: 293–299 (1975).

61 Bilheimer, D.W.; Goldstein, J.L.; Grundy, S.M., and Brown, M.S.: Reduction in cholesterol and low density lipoprotein synthesis after portacaval shunt surgery in a patient with homozygous familial hypercholesterolemia. J.clin.Invest. 56: 1420–1430 (1975).

62 Simons, L.A.; Reichl, D.; Myant, N.B., and Mancini, M.: The metabolism of the apoprotein of plasma low density lipoprotein in familial hyperbetalipoproteinaemia in the homozygous form. Atherosclerosis 21: 283–298 (1975).

63 Murthy, V.K.; Monchesky, T.C., and Steiner, G.: In vitro labeling of β-apolipoprotein with ^{3}H or ^{14}C and preliminary application to turnover studies. J.Lipid Res. 16: 1–6 (1975).

64 Packard, C.J.; Third, J.L.H.C.; Shepherd, J.; Lorimer, A.R.; Morgan, H.G., and Lawrie, T.D.V.: Low density lipoprotein metabolism in a family of familial hypercholesterolemic patients. Metabolism 25: 995–1006 (1976).

65 Reichl, D.; Simons, L.A.; Myant, N.B.; Pflug, J.J., and Mills, G.L.: The lipids and lipoproteins of human peripheral lymph, with observations on the transport of cholesterol from plasma and tissues into lymph. Clin.Sci.mol.Med. 45: 313–329 (1973).

66 Reichl, D.; Postiglione, A.; Myant, N.B.; Pflug, J.J., and Press, M.: Observations on the passage of apoproteins from plasma lipoproteins into peripheral lymph in two men. Clin.Sci.mol.Med. 49: 419–426 (1975).

67 Weinstein, D.B.; Carew, T.E., and Steinberg, D.: Uptake and degradation of low density lipoprotein by swine arterial smooth muscle cells with inhibition of cholesterol biosynthesis. Biochim.biophys. Acta 424: 404–421 (1976).

68 Balasubramaniam, S.; Goldstein, J.L.; Faust, J.R., and Brown, M.S.: Evidence for regulation of 3-hydroxy-3-methylglutaryl coenzyme A reductase activity and cholesterol synthesis in nonhepatic tissues of rat. Proc.natn.Acad.Sci. USA 73: 2564–2568 (1976).

69 Anderson, J.M. and Dietschy, J.M.: Cholesterogenesis: derepression in extrahepatic tissues with 4-aminopyrazolo[3,4-d]pyridine. Science 143: 903–905 (1976).

70 Glomset, J.A.: The plasma lecithin:cholesterol acyltransferase reaction. J.Lipid Res. 9: 155–167 (1968).

71 Skipski, V.P.; Barclay, M.; Barclay, R.K.; Fetzer, V.A.; Good, J.J., and Archibald, F.M.: Lipid composition of human serum lipoproteins. Biochem.J. *104:* 340–352 (1967).
72 Alaupovic, P.; Shafeek, S.; Furman, R.H.; Sullivan, M.L., and Walraven, S.L.: Studies of the composition and structure of serum lipoproteins. Isolation and characterization of very high density lipoproteins of human serum. Biochemistry, N.Y. *5:* 4044–4053 (1966).
73 Weinstein, D.B.; Carew, T.E., and Steinberg, D.: Uptake of low density lipoprotein by porcine aortic smooth muscle cells with inhibition of cholesterol synthesis. Circulation *50:* suppl. III, p. 70 (1974).
74 Steinberg, D.; Carew, T.E.; Weinstein, D.B., and Koschinsky, T.: Binding, uptake and catabolism of LDL and HDL by cultured smooth muscle cells; in Greten, Lipoprotein metabolism, pp. 90–98 (Springer, Heidelberg 1976).
75 Brown, M.S.; Faust, J.R., and Goldstein, J.L.: Role of the low density lipoprotein receptor in regulating the content of free and esterified cholesterol in human fibroblasts. J.clin.Invest. *55:* 783–793 (1975).
76 Stein, O.; Vanderhoek, J., and Stein, Y.: Cholesterol content and sterol synthesis in human skin fibroblasts and rat aortic smooth muscle cells exposed to lipoprotein-depleted serum and high density apolipoprotein/phospholipid mixtures. Biochim. biophys. Acta *431:* 347–358 (1976).
77 Stein, O. and Stein, Y.: The removal of cholesterol from Landschutz ascites cells by high density apolipoprotein. Biochim. biophys. Acta *326:* 232–244 (1973).
78 Stein, Y.; Glangeaud, M.C.; Fainaru, M., and Stein, O.: The removal of cholesterol from aortic smooth muscle cells in culture and Landschutz ascites cells by fractions of human high density apolipoproteins. Biochem. biophys. Acta *380:* 106–118 (1975).
79 Carew, T.E.; Koschinsky, T.; Hayes, S.B., and Steinberg, D.: A mechanism by which high density lipoproteins may slow the atherogenic process. Lancet *i:* 1315–1317 (1976).
80 Miller, N.E.; Weinstein, D.B.; Carew, T.E.; Koschinsky, T., and Steinberg, D.: Interaction between high density and low density lipoproteins during uptake and degradation by cultured human fibroblasts. J.clin.Invest. *60:* 78–88 (1977).
81 Hamilton, R.L.: Synthesis and secretion of plasma lipoproteins. Adv.exp.Med.Biol. *26:* 7–24 (1972).
82 Herbert, P.N.; Gotto, A.M., and Fredrickson, D.S.: Familial lipoprotein deficiency (abetalipoproteinemia, hypobetalipoproteinemia, and Tangier disease); in Stanbury, Wyngaarden and Fredrickson, Metabolic basis of inherited diseases; 4th ed., chap. 28, pp. 544–588 (McGraw-Hill, New York 1978).
83 Schaefer, E.J. and Brewer, H.B., jr.: Tangier disease: a defect in the conversion of chylomicrons to high-density lipoproteins. Clin.Res. *26:* 532A (1978).
84 Mahley, R.W. and Innerarity, T.L.: Interaction of canine and swine lipoproteins with the low density lipoprotein receptor of fibroblasts as correlated with heparin/manganese precipitability. J.biol.Chem. *252:* 3980–3986 (1977).
85 Sniderman, A.; Thomas, D.; Marpole, D., and Teng, B.: Low density lipoprotein – A metabolic pathway for return of cholesterol to the splanchnic bed. J.clin.Invest. *61:* 867–873 (1978).
86 Entenman, C.; Chaikoff, I.L., and Zilversmit, D.B.: Removal of plasma phospholipides as a function of the liver: the effect of exclusion of the liver on the turnover rate

of plasma phospholipides as measured with radioactive phosphorus. J. biol. Chem. *166:* 15–23 (1946).
87 Zilversmit, D.B. and Bollman, J.L.: Role of the liver and intestine in the turnover of plasma phosphatides in the rat. Archs Biochem. Biophys. *63:* 64–72 (1956).
88 Hotta, S. and Chaikoff, I.L.: The role of the liver in the turnover of plasma cholesterol. Archs Biochem. Biophys. *55:* 28–37 (1954).
89 Steinberg, D.: Lipoprotein metabolism – New insights from cell biology; Proc. 6th Int. Symp. on Drugs Affecting Lipid Metabolism; in Kritchevsky, Paoletti and Holmes, Advances in experimental medicine and biology; *109:* 3–27 (Plenum Press, New York 1978).
90 Brown, M.S. and Goldstein, J.L.: Regulation of the activity of the low density lipoprotein receptor in human fibroblasts. Cell *6:* 307–316 (1975).
91 Stein, Y.; Ebin, V.; Bar-on, H., and Stein, O.: Chloroquine-induced interference with degradation of serum lipoproteins in rat liver, studied *in vivo* and *in vitro*. Biochim. biophys. Acta *486:* 286–297 (1977).
92 Pittman, R.C. and Steinberg, D.: A new approach for assessing cumulative lysosomal degradation of proteins or other macromolecules. Biochem. biophys. Res. Commun. *81:* 1254–1259 (1978).
93 Cohn, Z.A. and Ehrenreich, B.A.: The uptake, storage and intracellular hydrolysis of carbohydrates by macrophages. J. exp. Med. *129:* 201–225 (1969).
94 Becker, G. and Ashwood-Smith. M.J.: Endocytosis in chinese hamster fibroblasts. Expl Cell Res. *82:* 310–314 (1973).
95 Kostner, G.M.: ApoB-deficiency (abetalipoproteinemia): a model for studying the lipoprotein metabolism; in Rommel and Bohmer, Lipid absorption: biochemical and clinical aspects, pp. 203–239 (University Park Press, Baltimore 1976).
96 Lees, R.S. and Ahrens, E.H.: Fat transport in abetalipoproteinemia: the effects of repeated infusions of β-lipoprotein-rich plasma. New Engl. J. Med. *280:* 1261–1270 (1969).
97 Levy, R.I.; Langer, T.; Gotto, A.M., and Fredrickson, D.S.: Familial hypobetalipoproteinemia, a defect in lipoprotein synthesis. Clin. Res. *18:* 539 (1970).
98 Sigurdsson, G.; Nicoll, A., and Lewis, B.: Turnover of apolipoprotein-B in two subjects with familial hypobetalipoproteinemia. Metabolism *26:* 25–31 (1977).
99 Steinberg, D.; Grundy, S.M.; Mok, H.Y.I.; Turner, J.D.; Weinstein, D.B.; Brown, W.V., and Albers, J.J.: Low density lipoprotein (LDL) and sterol metabolism in an unusual case of familial hypobetalipoproteinemia. Circulation *56:* suppl. III, p.4 (1977).
100 Goldstein, J.L. and Brown, M.S.: Binding and degradation of low density lipoproteins by cultured human fibroblasts. J. biol. Chem. *249:* 5153–5162 (1974).
101 Goldstein, J.L.; Brown, M.S., and Stone, N.J.: Genetics of the LDL receptor: evidence that the mutations affecting the binding and internalization are allelic. Cell *12:* 629–641 (1977).
102 Fogelman, A.M.; Seager, J.; Edwards, P.A., and Popjak, G.: Mechanism of induction of 3-hydroxy-3-methylglutaryl coenzyme A reductase in human leukocytes. J. biol. Chem. *252:* 644–651 (1977).
103 Haust, M.D.; More, R.H., and Movat, H.Z.: The role of smooth muscle cells in the fibrogenesis of arteriosclerosis. Am. J. Path. *37:* 377–389 (1960).
104 Fischer-Dzoga, K. and Wissler, R.W.: Stimulation of proliferation in stationary

primary cultures of monkey aortic smooth muscle cells. 2. Effect of varying concentrations of hyperlipemic serum and low density lipoproteins of varying dietary fat origins. Atherosclerosis *24:* 515–525 (1976).

105 Ross, R. and Harker, L. A.: Hyperlipidemia and arteriosclerosis: chronic hyperlipemia induces lesion initiation and progression by endothelial cell desquamation and lipid accumulation. Science *193:* 1094–1100 (1976).

106 Ross, R.; Glomset, J.; Kariya, B., and Harker, L.: A platelet-dependent serum factor that stimulates the proliferation of arterial smooth muscle cells *in vitro*. Proc. natn. Acad. Sci. USA *71:* 1207–1210 (1974).

107 Ross, R. and Glomset, J. A.: The pathogenesis of atherosclerosis. New Engl. J. Med. *295:* 420–425 (1976).

108 De Duve, C.: The participation of lysosomes in the transformation of smooth muscle cells to foamy cells in the aorta of cholesterol-fed rabbits. Acta cardiol. *20:* suppl. 1, pp. 9–25 (1974).

109 Goldstein, J. L. and Brown, M. S.: Lipoprotein receptors, cholesterol metabolism and atherosclerosis. Archs Path. *99:* 181–184 (1975).

110 Miller, N. E.; Weinstein, D. B., and Steinberg, D.: Uptake and degradation of high density lipoprotein: comparison of fibroblasts from normal subjects and from homozygous familial hypercholesterolemic subjects. J. Lipid Res. *19:* 644–653 (1978).

111 Stein, O.; Weinstein, D. B.; Stein, Y., and Steinberg, D.: Binding, internalization and degradation of low density lipoprotein by normal human fibroblasts and by fibroblasts from a case of homozygous familial hypercholesterolemia. Proc. natn. Acad. Sci. USA *73:* 14–18 (1976).

112 Brown, M. S. and Goldstein, J. L.: Familial hypercholesterolemia: a genetic defect in the low density lipoprotein receptor. New Engl. J. Med. *294:* 1386–1390 (1976).

113 Moore, A. T.; Williams, K. E., and Lloyd, J. B.: The effect of chemical treatment of [^{125}I]iodinated bovine serum albumin on its rate of pinocytosis by 17.5-day rat yolk-sac cultured *in vitro*. Biochem. Soc. Trans. *2:* 648–650 (1974).

114 Carew, T. E.; Chapman, J., and Steinberg, D.: Enhanced degradation of proteolytically damaged low density lipoprotein (LDL) by fibroblasts from patients with homozygous familial hypercholesterolemia. Circulation *56:* suppl. III, p. 99 (1977).

115 Basu, S. K.; Goldstein, J. L.; Anderson, R. G. W., and Brown, M. S.: Degradation of cationized low density lipoprotein and regulation of cholesterol metabolism in homozygous familial hypercholesterolemia fibroblasts. Proc. natn. Acad. Sci. USA *73:* 3178–3182 (1976).

116 Smith, E. B. and Smith, R. H.: Early changes in aortic intima; in Paoletti and Gotto, Atherosclerosis reviews, vol. 1, pp. 119–136 (Raven Press, New York 1976).

117 Miller, G. J. and Miller, N. E.: Plasma high density lipoprotein concentration and development of ischaemic heart disease. Lancet *i:* 16–19 (1975).

118 Castelli, W. P.; Doyle, J. T.; Gordon, T.; Hames, C.; Hulley, S. B.; Kagan, A.; McGee, D.; Vicic, W., and Zukel, W. J.: HDL cholesterol levels (HDLC) in coronary heart disease (CHD): a cooperative lipoprotein phenotyping study. Circulation *52:* II, p. 97 (1975).

119 Rhoads, G. G.; Gulbrandsen, C. L., and Kagan, A.: Serum lipoproteins and coronary heart disease in a population study of Hawaii Japanese man. New Engl. J. Med. *294:* 293–298 (1976).

120 Miller, N. E.; Forde, O. H.; Thele, D. S., and Mjøs, O. D.: The Tromsø heart study.

High density lipoprotein and coronary heart disease: a prospective case-control study. Lancet *i:* 965–967 (1977).
121 Reckless, J.P.D.; Weinstein, D.B., and Steinberg, D.: Lipoprotein metabolism by rabbit arterial endothelial cells. Circulation *54:* suppl. III, p. 56 (1976).
122 Reckless, J.P.D.; Weinstein, D.B., and Steinberg, D.: Lipoprotein and cholesterol metabolism in rabbit arterial endothelial cells in culture. Biochim, biophys. Acta *529:* 475–487 (1978).
123 Stein, O. and Stein, Y.: High density lipoproteins reduce the uptake of low density lipoproteins by human endothelial cells in culture. Biochim. biophys. Acta *431:* 363–368 (1976).

D. Steinberg, MD, PhD, Department of Medicine M-013,
University of California San Diego, La Jolla, CA 92093 (USA)

Composition and Metabolism of High-Density Lipoproteins

Ernst J. Schaefer and Robert I. Levy

National Heart, Lung and Blood Institute, National Institutes of Health, Bethesda, Md.

Introduction

Interest in high-density lipoproteins (HDL) has increased greatly in recent years, largely due to the finding that HDL cholesterol is an independent negative risk factor for coronary artery disease. Epidemiologic studies in Hawaii [1], Framingham (USA) [2], Norway [3] and Israel [4] all support this concept, originally suggested by the work of *Gofman et al.* [5] and *Barr et al.* [6]. Subjects with high HDL cholesterol levels appear to have increased longevity [7]. It has been postulated that HDL may be important in removing cholesterol from tissues, thereby reducing the amount of lipid deposition in the arterial wall [8].

Composition and Physical Properties

Lipoproteins

HDL, as isolated in the density range 1.063–1.21 g/ml by ultracentrifugation, are composed (weight %) of approximately 50% protein, 25% phospholipid, 20% cholesterol and 5% triglyceride [9–12]. The ratio of esterified:free cholesterol is about 3:1 and the lecithin:sphingomyelin ratio is 5:1 within HDL. Total HDL levels in human plasma are around 300 mg%. HDL has been divided into two density classes: HDL_2 (density range 1.063–1.125 g/ml) and HDL_3 (density range 1.125–1.21 g/ml). HDL_2 is composed of about 60% lipid and 40% protein, while 55% of the HDL_3 mass is due to protein. The lecithin:sphingomyelin ratio and the ratio of free cholesterol:esterified cholesterol are higher in HDL_2 than in HDL_3 [10–13]. Recently, *Anderson et al.* [14] have reported the presence of three HDL subfractions isolated by density gradient ultracentrifugation within the HDL density range. These subfractions have particle size ranges of 8.5–9.6, 9.7–10.7 and 10.8–12.8 nm, and two of these subfractions (HDL_{2a} and HDL_{2b}) are found in HDL_2. A negative correlation between HDL_2 and very-low-density lipoproteins (VLDL) has been noted [15]. Population studies suggest that fluctuations in HDL levels are largely due to changes in HDL_2 levels [16].

Various models for HDL structure have been proposed, based on data from chemical, enzymatic and electron microscopy studies, as well as nuclear magnetic resonance spectroscopy [17–19]. It appears that HDL particles are spherical micelles, approximately 90–120 Å in diameter, consisting of lipids with polar groups of the phospholipids at the surface or aqueous interface and with the globular apolipoproteins A (apoA) partially embedded in lipid [19]. The major forces in the protein-lipid interaction are apparently hydrophobic and protein-protein interactions may be of special importance to the organization of the native HDL molecule. The protein constituents of HDL include apolipoproteins A-I, A-II, B, C-I, C-II, C-III, D and E [12, 20, 21]. ApoA-I and apoA-II constitute about 90% of total HDL protein, with an apoA-I: apoA-II weight ratio of about 3:1 [10]. Published reports of the apoA-I: apoA-II ratio in HDL_2 compared to HDL_3 are contradictory, indicating variously that it is higher [12, 13, 22, 23], lower [24] or identical [25] to HDL_3.

The existence of several different lipoprotein particles within HDL has been documented, with each lipoprotein species containing different apolipoproteins. The presence of lipoprotein 'families' LP-A, LP-B, LP-C, and LP-D has been reported within HDL, and the composition of these lipoprotein species is shown in table I [12, 21]. LP-B was isolated using affinity chromatography, while LP-A was separated from LP-C by hydroxyapatite column chromatography. Whether this latter separation can be accomplished by immunological techniques remains to be determined. LP-C contains apoC-I, apoC-II and apoC-III as its protein constituents. Two types of LP-A species have been isolated, one in HDL_2 and one in HDL_3. Both contained apoA-I and apoA-II and were free of other apolipoproteins. LP-A is apparently the major lipoprotein within HDL. Whether LP-C is clearly a distinct species or is associated with LP-A remains to be firmly established. In addition, *Kostner et al.* [11] have reported the presence of an HDL subfraction within HDL_2 containing only apoA-I, while *Assman et al.* [26] have isolated an HDL particle from patients with Tangier disease containing apoA-II as its only protein constituent.

Table I. Composition (weight%) of reported lipoprotein species in HDL [12, 21]

Lipoprotein species	Protein	Cholesterol ester	Free cholesterol	Triglyceride	Phospholipid
	%	%	%	%	%
LP-A (HDL$_2$) apoA-I, apoA-II	40.1	13.1	4.4	3.4	32.2
LP-A (HDL$_3$) apoA-I, apoA-II	56.1	9.4	2.2	3.2	27.0
LP-B (HDL$_2$) apoB	39.5	25.1	6.8	5.1	21.0
LP-C (HDL) apoC-I, apoC-II, apoC-III	51.0	15.2	1.5	5.4	26.0
LP-D (HDL$_3$) apoD	70.0	5.7	5.7	3.3	15.3

Minor constituents within HDL include LP-D, LP-E, LP-F and LP (a) or sinking pre-β-lipoprotein. LP-D is a lipoprotein species containing only apoD, and has been isolated from HDL$_3$ [21]. Recently, *Curry et al.* [20] have reported the presence of LP-E, a lipoprotein species containing only apoE, found within HDL$_2$. The composition of this lipoprotein has not yet been reported. *Mahley et al.* [27] have isolated a cholesterol-rich HDL fraction (HDL$_c$) containing significant amounts of apoE from cholesterol-fed animals. HDL$_c$ may be similar to LP-E.

LP (a) or sinking pre-β-lipoprotein is found in varying amounts in the density range 1.055–1.085 g/ml. This lipoprotein is composed of 27% protein, 65% lipid and 8% carbohydrate. The protein moiety is composed of 65% apoB, 20% 'LP (a)' apolipoprotein and possibly albumin [28]. The major lipids are cholesterol (mainly esterified) and phospholipids. Recently, the LP (a) antigen was identified in 92% of 340 free-living adults, supporting the concept that it may exist in all individuals, although often only in small amounts [29]. This lipoprotein is apparently distinct from LP-B [12].

The data cited above are consistent with the concept that several different lipoprotein species exist within HDL, the major one being LP-A. Immunologic studies have helped to clarify the situation. We are still hampered by isolation techniques that may dissociate lipoprotein particles and alter their composition and buoyant density. Future work will hopefully clearly define the composition and metabolism of these lipoproteins.

Apolipoproteins

The major HDL apolipoproteins, apo A-I and apo A-II, can be readily isolated from human HDL, and are easily separated by various chromatographic techniques [9]. ApoA-I is largely helical in structure. Glutamine is the carboxy terminal amino acid, aspartic acid is the amino terminal, and isoleucine is absent. ApoA-I has been reported to activate lecithin:cholesterol acyl transferase (LCAT) [30]. The protein is a single polypeptide chain of 243 amino acid residues, and its amino acid sequence as reported by *Brewer et al.* [31] differs in several positions from earlier reports by *Baker et al.* [32, 33] and *Delahunty et al.* [34]. The reason for these differences remains unclear.

Several groups have reported that apoA-I can be fractionated into several distinct forms on DEAE ion-exchange chromatography [35, 36]. Most of this heterogeneity can be eliminated by omitting steps in the purification scheme known to cause aggregation (such as lyophilization or concentration by dialysis). Two polymorphic forms of apoA-I are still isolated utilizing these latter techniques [37]. These forms do not differ in their molecular weight (27,000 daltons), amino acid composition, and immunologic properties. The difference between these two polymorphic forms remains unknown [38].

ApoA-II is a single protein of molecular weight 17,000 daltons. By reduction of a single disulfide bond, two identical peptides of molecular weight 8,500 daltons are isolated [39, 40]. These peptides consist of 77 amino acid residues, with glutamine as the carboxy terminal amino acid and pyrolidone carboxylic acid as the amino terminal residue. These peptides lack histidine, arginine and tryptophan. The amino acid sequence of apoA-II has been reported [39, 40].

Lipid-protein interactions in plasma lipoproteins have been recently reviewed [41, 42]. Both apoA-I and apoA-II bind phosphatidylcholine and form protein-phospholipid complexes [43]. These complexes contain about 40% protein by weight, are isolated in the HDL density range 1.063–1.21 g/ml and can incorporate cholesterol ester. When either apolipoprotein is complexed with lipid, there is an increase in secondary structure, indicating that lipid-protein interactions may increase apoA secondary structure. However, both apoA-I and apoA-II self-associate in aqueous solutions, and at least for apoA-II, self-association is accompanied by major changes in secondary structure [38, 44, 45]. These data indicate that changes in secondary structure seen with apoA lipid recombination studies may be due to protein-protein interactions alone.

In normal plasma, 87% of apoA-I and 90% of apoA-II are found in HDL, and trace amounts of each apolipoprotein are found in other lipoprotein fractions [23]. 12% of apoA-I and 9% of apoA-II are found in the lipoprotein-free fraction (1.21 g/ml infranatant). The amount of apoA found in the lipoprotein-free fraction can be enhanced by repeated ultracentrifugation. Apparently, as much as 50% of immunoassayable apoA-I can be dissociated from HDL with repeated ultracentrifugations [46].

Plasma levels of apoA-I and apoA-II have been reported by several groups, utilizing either radioimmunoassay, electroimmunoassay or radial immunodiffusion techniques. Reported mean normal values range from 100 to 154 mg% for apoA-I and 34 to 83 mg% for apoA-II [22, 23, 46–49]. Differences in normal values may relate to whether or not the assay technique incorporated a delipidation step and possible variability in antisera.

The physical properties, functions and metabolism of the apoC (apoC-I, apoC-II and apoC-III) found in VLDL and HDL will be discussed in the chapter on VLDL metabolism. These apolipoproteins appear to play an impressive role as activators of various enzymes involved in lipoprotein metabolism. In fasting plasma the apoC are mainly found in VLDL and HDL, and constitute about 5% of the HDL protein mass [9]. In hypertriglyceridemic subjects, more apoC-II is found in VLDL and less in HDL than in normal subjects [50]. Specific immunoassays for the measurement of apoC levels in plasma have recently been developed and should help to elucidate apoC metabolism [50, 51].

ApoD, also known as 'thin-line peptide', has been isolated from HDL$_3$ and is a glycoprotein of molecular weight 22,700 daltons [21]. It is a minor protein constituent of HDL. *Kostner et al.* [52] have isolated an apolipoprotein from HDL also designated as 'thin-line peptide' or apoA-III which differs from apoD in that cystine is absent in this protein. ApoA-III has been reported to activate LCAT [53]. Recently, *Olofsson et al.* [54] have reported the presence of a new apolipoprotein (F) within HDL, of molecular weight 26,000–30,000 daltons. This acidic protein migrates as a single band on polyacrylamide-gel electrophoresis with a mobility similar to that of apoD and has an isoelectric focusing point of 3.7. This protein appears to carry its own complement of lipid and form a distinct lipoprotein within HDL. Further work is required to determine whether apoA-III, apoD, and apoF are all distinct entities.

ApoE or the arginine-rich protein was originally isolated from VLDL by *Shore et al.* [55], and shown to have a molecular weight of 33,000 daltons by *Shelburne et al.* [56]. *Utermann et al.* [57] have reported that several polymorphic forms of apoE may be present, with pI values of 5.5, 5.6, and 5.75 on isoelectric focusing, and that patients with type III hyperlipoproteinemia are deficient in one of these components (E_3). Recently, *Mahley et al.* [58] have isolated an apolipoprotein of molecular weight 48,000 daltons from the VLDL of type III hyperlipoproteinemia subjects and from the HDL_2 of normal subjects. Treatment of this protein, designated pro-arginine-rich protein (pro-ARP), with disulfide-reducing agents results in two subunits: a 37,000 molecular weight subunit which is immunologically and compositionally identical to apoE and a 10,000 molecular weight subunit. Whether pro-ARP is a precursor for apoE, a polymorphic form of apoE or a protein aggregate remains to be defined. ApoE is readily removed from both VLDL and HDL with ultracentrifugation [59]. Several groups have developed immunoassays for measuring plasma apoE levels. *Kushwaha et al.* [60] have suggested that elevated plasma apoE levels are diagnostic of type III hyperlipoproteinemia, while *Curry et al.* [20] have reported elevated apoE plasma levels in subjects with both type III and type V hyperlipoproteinemia. ApoE is a minor protein constituent of HDL.

Most of HDL appears to be made up of LP-A, containing apoA-I and apoA-II. ApoA-I and apoA-II plasma levels are 30% of normal in patients with LCAT deficiency and less than 25% of normal in hyperchylomicronemic subjects [23]. Patients with familial HDL deficiency (Tangier disease) have apoA-I plasma levels that are 1% or less of normal, while apoA-II levels are around 7% of normal [23, 26, 61]. Most of the apoA-II in these patients is found in the 1.063 g/ml supernatant [61] while most of the apoA-I is found in the 1.21 g/ml infranatant [23, 26]. Whether this abnormal apoA-I distribution represents the true situation or is due to isolation procedures remains to be determined.

HDL Cholesterol Levels

Plasma levels of apoA-I and apoA-II correlate with HDL cholesterol levels [22]. Fasting HDL cholesterol levels in normal and dyslipoproteinemic subjects studied at the National Institutes of Health (USA) while on *ad libitum* diets are shown in table II. Patients with all forms of primary hyperlipoproteinemia except type II, have HDL cholesterol levels that are significantly ($p < 0.001$) less than those found in normal subjects. In all normal and hyperlipoproteinemic subjects, HDL cholesterol is negatively correlated with VLDL cholesterol and plasma triglyceride levels. These correlations are statistically

Table II. Plasma lipid and HDL cholesterol levels in normal and dyslipoproteinemic subjects

	Plasma Cholesterol mg%	Plasma triglyceride mg%	HDL cholesterol mg%
Normal (n=1088)	189 ± 40	87 ± 43	50 ± 14
Type I (n=15)	306 ± 188	3106 ± 2455	16 ± 6
Type II (n=522)	432 ± 96	127 ± 83	44 ± 12
Type III (n=71)	432 ± 157	679 ± 485	38 ± 19
Type IV (n=286)	242 ± 62	385 ± 388	37 ± 11
Type V (n=107)	365 ± 189	1948 ± 2009	27 ± 12
Tangier (n=7)	68 ± 19	140 ± 73	2 ± 2

Results are mean values ± 1 SD.

significant (p < 0.001) utilizing either Pearson or Spearman correlation coefficient analysis [62]. Patients with the lowest levels of HDL cholesterol (type I, type V, Tangier disease), all have chylomicrons in the fasting state, suggesting a relationship between chylomicron and HDL (LP-A) metabolism.

When patients with type IV or V hyperlipoproteinemia are treated with diet or nicotinic acid, plasma triglyceride levels fall and HDL cholesterol levels rise [63]. HDL cholesterol levels can be decreased in normal subjects by placing them on high carbohydrate diets [64], and can be increased by estrogen [65] or nicotinic acid administration [66]. HDL cholesterol levels are elevated in premenopausal women [67] and long-distance runners [68]. An adequate understanding of the causes for these variations in HDL cholesterol and apoA plasma levels can only come after HDL metabolism has been elucidated.

Metabolism

Synthesis

The site of HDL (LP-A, LP-B, LP-C, LP-E and LP-D) synthesis in man remains to be determined. Rat liver perfusion studies indicate that the liver secretes particles in the HDL density range which differ from plasma HDL in that they are richer in apoE and poorer in apoA-I [69, 70]. Incorporation of ^{14}C-labeled amino acids into HDL apolipoproteins by perfused rat liver was twelve times higher for apoE than for apoA-I when measured as cpm/mg protein [69]. More recent rat liver perfusion data indicate that the liver can produce a nascent discoidal HDL particle, which has an apoE:apoA-I ratio of 10:1; while the apoE:apoA-I ratio in spherical plasma HDL was 1:7 [71]. Apparently, the rat liver can also produce an LP-B particle, found in the HDL_2 density range [72].

In patients with alcoholic hepatitis and LCAT deficiency, HDL levels are decreased and apoE constitutes about 40% of HDL protein mass [73]. No apoE is found in the VLDL of these patients. When plasma from these patients is incubated with LCAT, apoE is transferred from HDL to VLDL. Similarly, after recovery from hepatitis, the following events occurred in these patients: (1) LCAT activity returned to normal, (2) normal amounts of apoE were found in VLDL, (3) the apoE content of HDL protein fell to 1% and (4) HDL levels returned to normal. These findings are consistent with the concept that the liver produces a nascent discoidal HDL particle containing mainly apoE (LP-E), as well as some LP-B found within HDL_2. The bulk of LP-A would appear to be derived from other sources.

The intestine may be the major source of the apoA. ApoA-I and apoA-II account for 12% or more of the protein content of human thoracic duct lymph chylomicrons [74, 75]. ApoA-I has been identified as the major apolipoprotein of rat mesenteric lymph chylomicrons [76]. After exposure to serum, the content of apoE and apoC in rat mesenteric lymph chylomicrons increases, while apoA-I content decreases [77]. Immunofluorescent studies indicate that rapid and marked increases in intestinal apoA-I synthesis occur during lipid absorption, and this has been confirmed by ^{3}H-leucine incorporation studies [76]. Intestinal perfusion data in the rat are consistent with the concept that the intestine can produce lipoprotein particles containing apoB and the apoA, but not the apoC [78]. These latter apolipoproteins are apparently added to the particles after synthesis. Recently, we have shown that human lymph chylomicron apoA-I and apoA-II can serve as precursors for plasma HDL apoA-I and apoA-II [75]. Following injection of ^{125}I-labeled lymph chylomicrons into plasma, over 90% of apoA-I and apoA-II radioactivity was recovered in HDL within 1 h following injection. The data presented are consistent with the concept that LP-A arises from the intestine, while LP-C and LP-E appear to be synthesized by the liver. It appears that both liver and intestine can synthesize LP-B. The relative contribution of the intestine and liver to LP-A synthesis remains to be established as does the exact nature of the forms in which LP-A enters plasma and what regulates its catabolism.

Catabolism

The plasma kinetics of radioiodinated HDL in man has been studied by a number of investigators [66, 79–81]. Reported plasma half-life values in normal subjects range from 3.3 to 5.8 days. The half-life of HDL apolipoproteins is similar if injected as apoHDL or as HDL [80]. When injected as pure protein, 90% of the activity is recovered in the HDL fraction, indicating that apoHDL readily associates with HDL in plasma. Recently, ^{125}I-HDL (density range 1.09–1.21 g/ml) kinetics were studied in normal subjects, and the specific activity of apoA-I and apoA-II in HDL was measured serially for 14 days. These studies demonstrated that apoA-I catabolism and apoA-II catabolism within HDL are similar [66]. The plasma kinetics of radioiodinated apoA-I and apoA-II have also been studied. *Shepherd et al.* [82] have reported that apoA-I was catabolized at a significantly faster rate than apoA-II in 10 normal subjects while we have recently found that these two apolipoproteins were catabolized at similar rates in 12 normal individuals [83]. The reason for these differences is unclear.

The plasma kinetics of radioiodinated HDL has also been studied in patients with decreased levels of HDL. HDL catabolism is enhanced in nephrotic patients [79] and in hypertriglyceridemic subjects, especially those with hyperchylomicronemia [80]. It is also increased in subjects on high-carbohydrate diets [66] and is markedly enhanced in patients with familial HDL deficiency (Tangier disease) [84]. It appears that changes in HDL catabolism may play a major role in regulating HDL levels in plasma.

A multicompartmental model for HDL-apoA metabolism, shown in figure 1, has been developed utilizing the SAAM computer program [85]. Modeling methodology will be discussed in another chapter. This model was generated from plasma and urine radioactivity data collected during ^{125}I-HDL kinetic studies [66]. The model consists of a plasma and a nonplasma compartment, with exchange between these compartments, and catabolism from both compartments. HDL kinetic studies analyzed using this model revealed that HDL protein synthesis was not affected in 4 normal subjects studied before and during high-carbohydrate feeding, and was only slightly decreased in 2 subjects studied before and during nicotinic acid administration [66]. With high-carbohydrate feeding, the rate of HDL protein catabolism from the plasma compartment rose by 39.1% and with nicotinic acid treatment, it fell by 42.2%. Changes in the rate of catabolism from the nonplasma compartment were generally opposite to changes in catabolism from the plasma compartment, suggesting a reciprocal regulation of these

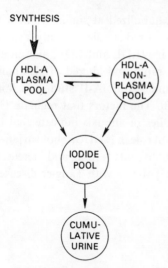

Fig. 1. Model for HDL-apoA metabolism.

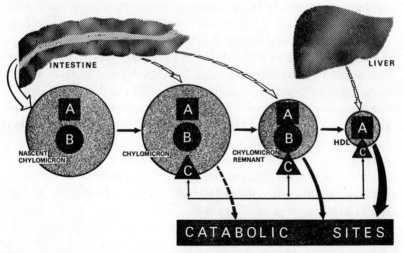

Fig. 2. Conceptual overview of A (apoA-I and ApoA-II) apolipoprotein metabolism.

two catabolic pathway. The exact nature of these pathways remains to be defined.

^{125}I HDL studies in patients with Tangier disease have shown that heterozygotes have normal HDL protein synthesis, while in homozygotes synthesis is about 50% of normal. HDL protein catabolism is twofold enhanced in heterozygotes and over tenfold enhanced in homozygotes. In heterozygotes, apoA-I and apoA-II are catabolized at similar rates within HDL, while in homozygotes apoA-I is catabolized at a significantly faster rate than apoA-II, and radioactivity is found in VLDL and LDL, as well as in HDL, indicating that homozygotes have rapid and altered HDL apolipoprotein catabolism [83, 84]. ^{125}I-apoA-I and ^{131}I-apoA-II studies in homozygotes have confirmed these findings [83]. The factors that regulate HDL catabolism remains to be elucidated. Studies in animals indicate that liver and kidney lysosomes may play an important role in HDL catabolism [86–88]. Why HDL catabolism is enhanced by carbohydrate feeding and increased in hyperchylomicronemic subjects as well as patients with Tangier disease remains to be determined.

Conclusion

Because HDL cholesterol has been shown to be a negative independent risk factor for coronary artery disease, interest in this lipoprotein density class has increased greatly. A number of different lipoprotein particles appear to exist within HDL, and further definition of these particles clearly is needed (LP-A, LP-B, LP-C, LP-D, LP-E, LP-F) [12, 20, 21, 54]. The major particle type within HDL appears to contain both apoA-I and apoA-II. Recent data suggest that LP-A may arise in the intestine [76] and that lymph chylomicron apoA can serve as precursors for plasma HDL apoA [75]. A conceptual overview of apoA metabolism is shown in figure 2. Patients with a defect in chylomicron metabolism would be expected to have low levels of HDL, as is shown in table II. Considerably more research is needed to clearly define the interrelationships between chylomicron and HDL metabolism. In the next decade much more information about HDL should become available, including its role as a risk factor in atherosclerosis.

References

1 Rhoads, G.; Gulbrandsen, C.L., and Kagan, A.: Serum lipoproteins and coronary heart disease in a population study of Hawaii Japanese men. New Engl. J. Med. *294:* 293–295 (1976).
2 Castelli, W.; Doyle, J.T.; Gordon, T.; Hames, C.G.; Hjortland, M.; Hulley, S.B; Kagan, A., and Zukel, W.J.: HDL cholesterol and other lipids in coronary heart disease, the cooperative lipoprotein phenotyping study. Circulation *55:* 767–772 (1977).

3 Miller, N.E.; Thelle, D.S.; Forde, O.H., and Mjos, O.D.: The Tromsö heart study, high density lipoprotein and coronary heart disease: a prospective case – control study. Lancet i: 965–967 (1977).
4 Medalie, J.H.; Kahn, H.A.; Neufeld, H.N.; Riss, E., and Goldbourt, U.: Five year myocardial infarction incidence. II. Association of single variables by age and birth place. J. chron. Dis. 26: 329–349 (1973).
5 Gofman, J.W.; Young, W., and Tandy, R.: Ischemic heart disease, atherosclerosis, and longevity. Circulation 34: 679–697 (1966).
6 Barr, D.P.; Russ, E.M., and Eder, H.A.: Protein lipid relationships in human plasma. II. In atherosclerosis and related conditions. Am. J. Med. 11: 480–493 (1951).
7 Glueck, C.J.; Fallat, R.W.; Millett, F.; Garside, P.; Elston, R.C., and Go, R.C.P.: Familial hyper-alpha-lipoproteinemia: studies in eighteen kindreds. Metabolism 24: 1243–1265 (1975).
8 Carew, T.E.; Hayes, S.B.; Kochinsky, T., and Steinberg, D.: A mechanism by which high density lipoproteins may slow the atherogenic process. Lancet i: 1315–1318 (1976).
9 Levy, R.I.; Blum, C.B., and Schaefer, E.J.: The composition, structure and metabolism of high density lipoproteins; in Lipoprotein metabolism, pp. 56–64 (Springer, Berlin 1976).
10 Scanu, A.M.; Lim, C.T., and Edelstein, C.: On the subunit structure of the protein of human serum high density lipoprotein. II. A study of Sephadex fraction IV. J. biol. Chem. 247: 5850–5855 (1972).
11 Kostner, G.M.; Patsch, J.R.; Sailer, S.; Braunsteiner, H., and Holasek, A.: Polypeptide distribution of main lipoprotein density classes separated from human plasma by rate zonal ultracentrifugation. Eur. J. Biochem. 15: 611–621 (1974).
12 Kostner, G.M. and Alaupovic, P.: Studies of the composition and structure of plasma lipoprotein, separation, and quantitification of lipoprotein families occurring in high density lipoproteins of human plasma. Biochemistry, N.Y. 11: 3419–3428 (1972).
13 Borut, T.C. and Aladjem, F.: Immunochemical heterogenity of serum high density lipoproteins. Immunochemistry 8: 851–863 (1971).
14 Anderson, D.W.; Nichols, A.V.; Forte, T.M., and Lindgren, F.T.: Particle distribution of human serum high density lipoproteins. Biochim. biophys. Acta 493: 55–68 (1977).
15 Nichols, A.V.: Human serum lipoproteins and their interrelationships; in Lawrence and Gofman, Advances in biological and medical physics; vol. II, pp. 110–156 (Academic Press, New York 1967).
16 Anderson, D.W.; Nichols, A.V.; Pan, S.S., and Lindgren, F.T.: Determination of three major components within serum high density distributions of a normal population sample. Atherosclerosis 29: 161–174 (1978).
17 Stoffel, W.; Zierenberg, O.; Tunggal, B., and Schreiber, E.: ^{13}C nuclear magnetic resonance spectroscopic evidence for hydrophobic lipid-protein interactions in human high density lipoproteins. Proc. natn. Acad. Sci. USA 71: 3696–3700 (1974).
18 Segrest, J.P.; Jackson, R.L.; Morrisett, J.D., and Gotto, A.M., jr.: A molecular theory of lipid-protein interactions in the plasma lipoproteins. FEBS Lett. 38: 247–253 (1974).
19 Assman, G. and Brewer, H.B., jr.: A molecular model of high density lipoproteins. Proc. natn. Acad. Sci. USA 71: 1534–1538 (1974).

20 Curry, M.D.; MC Conathy, W.J.; Alaupovic, P.; Ledford, J.H., and Popovic, M.: Determination of human apolipoprotein E by electroimmunoassay. Biochim. biopys. Acta 439: 413–425 (1976).

21 McConathy, W.J. and Alaupovic, P.: Studies on the isolation and partial characterization of apolipoprotein D and lipoprotein D in human plasma. Biochemistry, N.Y. 15: 515–520 (1976).

22 Cheung, M.C. and Albers, J.J.: The measurement of apolipoprotein A-I and A-II levels in men and women by immunoassay. J. clin. Invest. 60: 43–50 (1977).

23 Curry, M.D.; Alaupovic, P., and Suenram, C.A.: Determination of apolipoprotein A and its constitutive A-I and A-II polypeptides by separate electroimmunoassays. Clin. Chem. 22: 315–322 (1976).

24 Albers, J.J. and Aladjem, F.: Precipitation of ^{125}I-labeled lipoproteins with specific polypeptide antisera. Evidence for two populations with differing polypeptide composition in human high density lipoproteins. Biochemistry, N.Y. 10: 3436–3442 (1971).

25 Friedberg, S. and Reynolds, J.A.: The molar ratio of the two major polypeptide components of human high density lipoproteins. J. biol. Chem. 251: 4005–4009 (1976).

26 Assman, G.; Herbert, P.N.; Fredrickson, D.S., and Forte, T.: Isolation and characterization of an abnormal high density lipoprotein in Tangier disease. J. clin. Invest. 60: 242–252 (1977).

27 Mahley, R.W.; Weisgraber, K.H.; Innerarity, T.; Brewer, H.B., jr., and Assman, G.: Swine lipoproteins and atherosclerosis: changes in the plasma lipoproteins and apoproteins induced by cholesterol feeding. Biochemistry, N.Y. 14: 2817–2823 (1975).

28 Ehnholm, C.; Garoff, O.; Renkonew, O., and Simms, K.: Protein and carbohydrate composition of Lp (a) lipoprotein from human plasma. Biochemistry, N.Y. 11: 3229–3232 (1972).

29 Albers, J.J.; Adolfson, J.L., and Hazzard, W.R.: Radioimmunoassay of human plasma Lp (a) lipoprotein. J. Lipid Res. 18: 331–335 (1977).

30 Fielding, C.J.; Shore, V.G., and Fielding, P.E.: A protein co-factor of lecithin: cholesterol acyltransferase. Biochem. biophys. Res. Commun. 46: 1493–1498 (1972).

31 Brewer, H.B., jr.; Fairwell, T.; Larue, A.; Ronan, R.; Houser, A., and Bronzert, T.: The amino acid sequence of human apoA-I, an apolipoprotein isolated from high density lipoproteins. Biochem. biophys. Res. Commun. 80: 623–630 (1979).

32 Baker, H.N.; Gotto, A.M., jr., and Jackson, R.L.: The primary structure of human plasma high density apolipoprotein glutamine I (apoA-I). II. The amino acid sequence and alignment of cyanogen bromide fragments IV, III and I. J. biol. Chem. 250: 2725–2738 (1975).

33 Baker, H.N.; Jackson, R.L., and Gotto, A.M., jr.: Isolation and characterization of the cyanogen bromide fragments from the high-density apolipoprotein glutamine I. Biochemistry, N.Y. 12: 3866–3871 (1973).

34 Delahunty, T.; Baker, H.N.; Gotto, A.M., jr., and Jackson, R.L.: The primary structure of human plasma high density apolipoprotein glutamine I (apoA-I). I. The amino acid sequence of cyanogen bromide fragment II. J. biol. Chem. 250: 2718–2724 (1975).

35 Lux, S.E. and John, K.M.: Further characterization of the polymorphic forms of a human high density apolipoprotein, apo LP-Gln-I (apoA-I). Biochim. biophys. Acta 278: 266–270 (1972).

36 Edelstein, C.; Lim, C.T., and Scanu, A.M.: On the subunit structure of human serum high density lipoprotein. I. A study of its major polypeptide component (Sephadex, fraction III). J. biol. Chem. 247: 5842–5849 (1972).
37 Houser, A.; Larue, A.; Fairwell, T.; Osborne, J.C., jr., and Brewer, H.B., jr.: manuscript in preparation (1978).
38 Osborne, J.C., jr. and Brewer, H.B., jr.: The plasma lipoproteins. Adv. Protein Chem. 31: 253–337 (1977).
39 Brewer, H.B., jr.; Lux, S.E.; Ronan, R., and John, K.M.: Amino acid sequence of human apoLP-Gln-II (apoA-II), an apolipoprotein isolated from the high-density lipoprotein complex. Proc. natn. Acad. Sci. USA 69: 1304–1380 (1972).
40 Jackson, R.L. and Gotto, A.M., jr.: A study of the cystine-containing apolipoprotein of human plasma high density lipoproteins: characterization of cyanogen bromide and tryptic fragments. Biochim. biophys. Acta 285: 36–47 (1972).
41 Jackson, R.L.; Morrisett, J.D., and Gotto, A.M., jr.: Lipoprotein structure and metabolism. Physiol. Rev. 56: 259–316 (1976).
42 Morrisett, J.D.; Jackson, R.L., and Gotto, A.M., jr.: Lipid-protein interactions in the plasma lipoproteins. Biochim. biophys. Acta 472: 93–133 (1977).
43 Lux, S.E.; Hirz, R.; Shrager, R.I., and Gotto, A.M., jr.: The influence of lipid on the conformation of human plasma high density apolipoproteins. J. biol. Chem. 247: 2598–2606 (1972).
44 Gwynne, J.H.; Brewer, H.B., jr., and Edelhoch, H.: The molecular behavior of apoA-I in human high density lipoprotein. J. biol. Chem. 250: 2269–2274 (1975).
45 Osborne, J.C.; Palumbo, G.; Brewer, H.B., jr., and Edelhoch, H.: The self-association of the reduced apoA-II apoprotein from the human high density lipoprotein complex. Biochemistry, N.Y. 14: 3741–3746 (1975).
46 Fainaru, M.; Glangeaud, M.C., and Eisenberg, S.: Radioimmunoassay of human high density lipoprotein apoprotein A-I. Biochim. biophys. Acta 386: 432–443 (1975).
47 Schonfeld, G. and Pfleger, B.: The structure of high density lipoproteins and the levels of apolipoprotein A-I in plasma as determined by radio-immunoassay. J. clin. Invest. 54: 236–246 (1974).
48 Karlin, J.B.; Juhn, D.J.; Starr, J.L.; Scanu, A.M., and Rubenstein, A.H.: Measurement of human high density lipoprotein apolipoprotein A-I in serum by radioimmunoassay. J. Lipid Res. 17: 30–37 (1976).
49 Schonfeld, G.; Chen, J.; MC Donnel, W.F., and Jeng, I.: Apolipoprotein A-II content of human plasma high density lipoproteins measured by radioimmunoassay. J. Lipids Res. 18: 645–655 (1977).
50 Kashyap, M.L.; Srivastava, L.S.; Chen, C.Y.; Perisutti, G.; Campbell, M.; Lutner, R.F., and Glueck, C.J.: Radioimmunoassay of human apolipoprotein C-II. J. clin. Invest. 60: 171–180 (1977).
51 Curry, M.D.; Alaupovic, P., and Mc Conathy, W.: Serum apolipoprotein patterns of normolipidemic subjects and patients with primary hyperlipidemias. Abstract. Circulation 54: suppl. II, p. 134 (1976).
52 Kostner, G.M.: Studies on the composition and structure of human serum lipoproteins. Isolation and partial characterization of apolipoprotein A-III. Biochim. biophys. Acta 336: 383–395 (1974).
53 Kostner, G.: Studies on the cofactor requirement for lecithin: cholesterol acyltransferase. Scand. J. clin. Lab. Invest. 33: suppl. 137, pp. 19–21 (1974).

54 Olofsson, S.; Mc Conathy, W., and Alaupovic, P.: Isolation and partial characterization of an acidic apolipoprotein from human high density lipoproteins. Circulation 56: suppl. III, p. 56 (1977).
55 Shore, B. and Shore, V.: An apolipoprotein preferential enriched in cholesterol ester-rich low density lipoproteins. Biochem. biophys. Res. Commun. 58: 107 (1974).
56 Shelburne, F. A. and Quarfordt, S. H.: A new apoprotein of human plasma very low density lipoproteins. J. biol. Chem. 249: 1428–1433 (1974).
57 Utermann, G. M.; Jaeschke, M., and Menzel, J.: Familial hyperlipoproteinemia type III: deficiency of a specific apolipoprotein (apoE-III) in the very low density lipoproteins. FEBS Lett. 56: 352–355 (1975).
58 Mahley, R. W.; Weisgraber, K. H.; Innerarity, T. L., and Bersot, T. P.: Identification of pro-arginine-rich apoprotein – a possible modulator of lipoprotein binding to cell surface receptors of human fibroblasts. Circulation 56: suppl. III, p. 57 (1977).
59 Fainaru, M.; Havel, R. J., and Imaizumi, K.: Apolipoprotein content of plasma lipoproteins of the rat separated by gel chromatography or ultracentrifugation. Biochem. Med. 17: 347–355 (1977).
60 Kushwaha, R. S.; Hazzard, W. R.; Wahl, P. W., and Hoover, J. J.: Type III hyperlipoproteinemia: diagnosis in whole plasma by apolipoprotein E immunoassay. Ann. intern. Med. 87: 509–516 (1977).
61 Alaupovic, P.; Schaefer, E. J., and Brewer, H. B., jr.: unpublished observations.
62 Pearl, R.: Introduction to medical biometry and statistics (Saunders, Philadelphia 1963).
63 Carlson, L. A.; Olsson, A. G., and Ballantyne, D.: On the rise in low density and high density lipoproteins in response to the treatment of hypertriglyceridemia in type IV and type V hyperpoproteinemias. Atherosclerosis 26: 603–609 (1977).
64 Levy, R. I.; Lees, R. S., and Fredrickson, D. S.: The nature of prebeta (very low density) lipoproteins. J. clin. Invest. 45: 63–77 (1966).
65 Schaefer, E. J.; Levy, R. I.; Jenkins, L. L., and Brewer, H. B., jr.: The effects of estrogen administration on human lipoprotein metabolism. Proc. 6th Int. Symp. on Drugs Affecting Lipid Metabolism, Philadelphia 1977.
66 Blum, C. B.; Levy, R. I.; Eisenberg, S.; Hall, M. H.; Goebel, R. H., and Berman, M.: High density lipoprotein metabolism in man. J. clin. Invest. 60: 795–807 (1977).
67 Fredrickson, D. S. and Levy, R. I.: Familial hyperlipidemia; in Stanbury, Wyngaarden and Fredrickson, The metabolic basis of inherited disease, pp. 545–833 (Mc Graw-Hill, New York 1972).
68 Wood, P. D.; Haskell, W.; Klein, H.; Lewis, S.; Stern, M. P., and Farquhar, J. W.: The distribution of plasma lipoproteins in middle-aged male runners. Metabolism 25: 1249–1257 (1976).
69 Marsh, J.: Apoproteins of the lipoproteins in a nonrecirculating perfusate of rat liver. J. Lipid Res. 17: 85–90 (1976).
70 Hamilton, R. L.; Williams, M. C.; Fielding, C. J., and Havel, R. J.: Discoidal bi-layer structure of nascent high density lipoproteins from perfused rat liver. J. clin. Invest. 58: 667–680 (1976).
71 Felker, T. E.; Fainaru, M.; Hamilton, R. L., and Havel, R. J.: Secretion of the arginine rich and A-I apolipoproteins by the isolated perfused rat liver. J. Lipid Res. 18: 465–473 (1977).

72 Fainaru, M.; Felker, T.E.; Hamilton, R.L., and Havel, R.J.: Evidence that a separate particle containing only B apoprotein is present in high-density lipoprotein from perfused rat liver. Metabolism 26: 999–1004 (1977).
73 Ragland, J.B.; Bertram, P.D., and Sabesin, S.: Identification of nascent high density lipoprotein containing arginine-rich protein in human plasma. Biochem. biophys. Res. Commun. 80: 81–88 (1978).
74 Kostner, G. and Holasek, A.: Characterization and quantification of the apolipoproteins from human chyle chylomicrons. Biochemistry, N.Y. 11: 1217–1223 (1972).
75 Schaefer, E.J.; Jenkins, L.L., and Brewer, H.B., jr.: Human chylomicron apolipoprotein metabolism. Biochem. biophys. Res. Commun. 80: 405–412 (1978).
76 Glickman, R.M. and Green, P.H.R.: The intestine as a source of apolipoprotein A-I. Proc. natn. Acad. Sci. USA. 74: 2569–2573 (1977).
77 Imaizumi, K.; Fainaru, M., and Havel, R.J.: Transfer of apolipoproteins (A-I and ARP) between rat mesenteric lymph chylomicrons and serum lipoproteins. Abstract. Circulation 54: suppl. II, p. 134 (1976).
78 Windmueller, H.G.; Herbert, P.N., and Levy, R.I.: Biosynthesis of lymph and plasma lipoprotein apoprotein by isolated perfused rat liver and intestine. J. Lipid. Res. 14: 215–223 (1973).
79 Gitlin, D.; Cornwell, D.G.; Nakasato, D.; Oncley, J.L.; Hughes, W.L., jr., and Jameway, C.A.: Studies on the metabolism of the plasma lipoproteins in the nephrotic syndrome. J. clin. Invest. 37: 172–184 (1958).
80 Furman, R.H.; Sanbar, S.S.; Alaupovic, P.; Bradford, R.H., and Howard, R.P.: Studies of the metabolism of radioiodinated human serum alpha-lipoprotein in normal and hyperlipidemic subjects. J. Lab. clin. Med. 63: 193–204 (1964).
81 Scanu, A. and Hughes, W.L.: Further characterization of the human serum d 1.063-1.21 lipoproteins. J. clin. Invest. 41: 1681–1689 (1962).
82 Shepherd, J.; Packard, C.; Patsch, J., and Gotto, A.M.: Metabolism of apoprotein A-I and A-II. Abstract. Circulation 56: suppl. III, p. 5 (1977).
83 Schaefer, E.J.; Zech, L.; Jenkins, L.L.; Aamodt, R., and Brewer, H.B., jr.: unpublished observations.
84 Schaefer, E.J.; Blum, C.B.; Levy, R.I.; Jenkins, L.L.; Alaupovic, P.; Foster, D.M., and Brewer, H.B., jr.: Metabolism of high density apolipoproteins in Tangier disease. New Engl. J. Med. 299: 905–910 (1978).
85 Berman, M. and Weiss, M.F.: SAAM manual. US Public Health Service, publ. No. 1703 (US Government Printing Office, Washington 1972).
86 Nakai, T. and Whayne, T.F.: Catabolism of canine apolipoprotein A-I, purification, catabolic rate, organs of catabolism and the liver subcellular catabolic site. J. Lab. clin. Med. 88: 63–80 (1976).
87 Eisenberg, S.; Windmueller, H.G., and Levy, R.I.: Metabolic fate of rat and human lipoproteins in the rat. J. Lipid Res. 14: 446–458 (1973).
88 Rachmilewitz, P.; Stein, O.; Roheim, P.S., and Stein, Y.: Metabolism of iodinated high density lipoproteins in the rat, autoradiographic localization in the liver. Biochim. biophys. Acta 270: 414–425 (1972).

E.J. Schaefer, MD, National Heart, Lung, and Blood Institute,
National Institutes of Health, Bethesda, MD 20014 (USA)

Catabolism of Serum Lipoproteins

O. Stein and Y. Stein

Lipid Research Laboratory, Department of Medicine B,
Hadassah University Hospital and Department of Experimental Medicine
and Cancer Research, Hebrew University, Hadassah Medical School, Jerusalem

Introduction

Catabolism of serum lipoproteins is a multistep process, the purpose of which is to deliver to peripheral tissues various lipid constituents synthesized in the liver or small intestine. This part of the catabolic process could be designated the functional part, in contradistinction to the customary implication of degradation for the purpose of disposal. The enzymes involved in the process are of two types, surface-bound ectoenzymes as well as lysosomal acid hydrolases. One of the products of lipoprotein degradation is free cholesterol which cannot be metabolized further and the efflux of which from peripheral cells requires a removal system. The latter is perhaps not connected directly to the catabolic process *per se,* but might utilize certain lipophilic peptides or phospholipid-apoprotein complexes liberated from lipoprotein linkages during the steps of extracellular degradation. This chapter was written with an emphasis on the functional aspects of lipoprotein catabolism and was based mainly on more recent data. Therefore, it is quite selective and not exhaustive and obviously some valuable contributions were not included.

Very-Low-Density Lipoproteins and Chylomicrons

Functional Significance of the Catabolic Process

The function of the triglyceride-rich lipoproteins of plasma (very-low-density lipoproteins, VLDL, and chylomicrons) begins with their degradation, a stepwise process that results in a continuous formation of free fatty acids, which are utilized for the production of energy in all muscular tissue, especially in the heart [1]. Free fatty acids are used also for cellular phospholipid synthesis and thus for membrane formation. Serum lipoproteins are the obligate precursors of tissue essential fatty acids which, in addition to their function in membrane synthesis, are the only source for synthesis of prostaglandins, the ubiquitous local hormones. The catabolism of VLDL and chylomicron triglyceride plays also a prominent role in the supply of free fatty acids for milk formation [1] and is mandatory for the entry of free fatty acids into storage organs, such as adipose tissue [1]. It appears also that, in the lactating mammary gland, the catabolism of VLDL and chylomicrons provides free cholesterol which is utilized for milk secretion [2].

Recent studies have provided evidence that the chylomicrons secreted by the intestine undergo prominent changes in the plasma prior to their degradation [3]. These consist in the acquisition of C apoproteins, which provide them with the activator of lipoprotein lipase (apoC-II) and the arginine-rich protein (ARP) [4], which perhaps serves as a signal for the removal of the remnants from the circulation.

Catabolism of Triglyceride Moiety of Chylomicrons and VLDL

The key enzyme in the catabolism of the triglyceride portion is lipoprotein lipase [1] and the process takes place apparently at the luminal surface of the capillary endothelium, and probably vascular endothelium in other sites. Morphological studies on mammary gland [5] and adipose tissue [6] have localized the site of lipoprotein lipase activity to the endothelial cell surface. Another experimental approach that helped to define lipoprotein lipase as an ectoenzyme was perfusion of hearts with concanavalin A which resulted in inhibition of enzyme release [7]. More recently, inhibition of chylomicron lipolysis was achieved by perfusion of hearts with an antibody to lipoprotein lipase [8]. In most organs, lipoprotein lipase is present in two compartments, the functional one located at the surface of cells which come into contact with circulating triglycerides, the other compartment is intracellular, the site of enzyme production and perhaps its storage [1]. This dual localization of the enzyme has prompted studies directed towards elucidation of the mechanism of transport of the enzyme. Studies with perfused heart have shown that the transport mechanism can be interrupted by colchicine [9] and these findings were corroborated later in adipose tissue [10]. Recently, rat heart mesenchymal cells in culture which synthesize the enzyme quite actively [11], were shown to retain lipoprotein lipase at the intracellular site after colchicine treatment [12]. The culture system provided also an opportunity to study further the results of an interaction between heparin and lipoprotein-lipase-producing cells. It became apparent that even after a brief exposure to heparin the cells continue to release the enzyme into the incubation medium. The release acted also as a stimulus for new enzyme synthesis which could be inhibited by cycloheximide treatment [12]. The cultured rat heart mesenchymal cells gave also further support that hydrolysis of VLDL triglyceride occurs at the cell surface [13]. In these experiments labeled VLDL triglycerides were added to the incubation medium containing excess albumin and their hydrolysis resulted in retention of the free fatty acids in the medium. No degradation of VLDL triglyceride occurred in the medium in the absence of cells and lipoprotein lipase was not released into the medium containing VLDL in the absence of heparin [13]. These studies with cultured cells as well as results obtained with perfused hearts provide evidence that the catabolism of VLDL takes place on the plasma membrane [14] and not in the circulation following interaction between enzyme and substrate. The major substrate of the enzyme, i.e. the triglyceride, is situated in the core of the particle and the exact sequence of events leading to the interaction between the enzyme and this substrate remains to be elucidated. Since lipoprotein lipase is able to catalyze also phospholipid hydrolysis [15], the degradation of the surface phospholipids could be the primary event prior to the actual hydrolysis of the triglyceride.

Fate of the Remnant Particle

The concept of remnant particles was introduced by *Redgrave* [16], who has shown that following injection of chylomicrons to hepatectomized rats, little or no cholesterol ester became cleared from the circulation, in the presence of considerable removal of triglyceride. Subsequent studies with VLDL labeled in the protein portion with ^{125}I have shown that during intravascular catabolism of injected VLDL a particle is formed, which can be isolated at a density of 1.019 g/ml and which contains the ^{125}I-labeled apoB derived from the original VLDL, and which has been designated 'intermediate particle' [17]. The composition of the intermediate particles and their parent VLDL has been determined after *in vitro* degradation of VLDL with lipoprotein lipase [18]. It appears that after the degradation of the triglyceride the particle retains all apoB, loses about 40% of ARP and more than 90% of apoC. It contains also only one third of the phospholipid and free cholesterol and 70% of the cholesterol ester present in the original VLDL [18].

In the rat, the liver was shown to play a major role in the clearance from the circulation of both chylomicron [19] and VLDL remnants [20] as evidenced by the recovery of 70–90% of the injected labeled cholesterol ester, 20–30 min after injection of the labeled lipoproteins. More than 80% of ^{125}I-

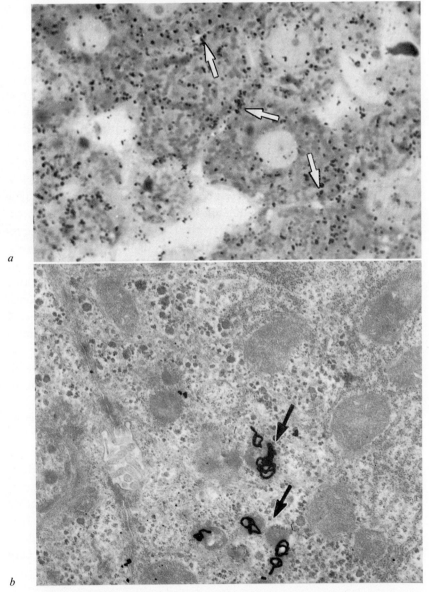

Fig.1. a Radioautograph of a section of rat liver, 120 min after injection of homologous ^{125}I-labeled VLDL. The radioautographic reaction which represents mainly apoB protein is seen over hepatocytes and is clustered in peribilliary regions (arrows). ×1,200. *b* Electron-microscopic radioautograph of some rat liver as in *a*. The silver grains are associated with secondary lysosomes. ×17,000. From *Stein et al.* [68].

apoB injected as a part of VLDL was cleared from the circulation within 60 min of injection and a major portion was recovered in the liver [21]. The uptake of the chylomicron [22] and VLDL remnants [23] by the liver was studied with the help of radioautography and the hepatocytes were shown to be the main site of remnant uptake. At early times after injection, a prominent radioautographic reaction was seen at the sinusoidal cell boundary indicating that binding to receptor sites might precede intracellular uptake [22, 23]. At later time intervals, the intracellular label was found concentrated over secondary lysosomes [23] (fig. 1). The role of hepatic acid hydrolases in the catabolism of VLDL remnant particles was elucidated further after injection of chloroquine which is a known inhibitor of cathepsin B_1 and which induced changes in hepatocyte lysosomes within a few hours of intraperitoneal injection [20]. Chloroquine administration did not affect the disappearance from the circulation of VLDL labeled with ^{125}I, or with ^3H-cholesterol linoleate. The retention of the ^{125}I-labeled protein was much higher in the livers of chloroquine-treated animals [20]. Since ^{125}I-labeled amino acids are not reutilized for protein synthesis, the finding of higher retention provided evidence that chloroquine administration resulted in impairment of hydrolysis of the protein moiety. Even though the total retention of label in the liver after injection of VLDL labeled with ^3H-cholesterol linoleate was similar in chloroquine-treated and in control livers, the degree of hydrolysis of cholesterol ester was markedly reduced in the experimental group [20]. Thus, these experiments provided also evidence that lysosomal cholesterol esterase is involved in the hydrolysis of remnant cholesterol ester.

Impairment in degradation of chylomicron remnant cholesterol ester by liver cells was found in response to colchicine treatment [24]. The explanation proposed for this effect was a possible impedance in fusion between the endocytic vacuole bearing the remnant particle and primary lysosomes. However, colchicine treatment does not interfere with the process of lysosome-phagosome fusion in macrophages [25]. In other studies, colchicine was shown to reduce markedly the heparin-releasable hepatic triglyceride hydrolase [26]. If this enzyme plays a role in the interiorization of the remnant particles, the reduction in hydrolysis of the cholesterol ester remnant could have been secondary to a fall in surface-located hepatic triglyceride hydrolase.

Fig. 2. Electron-microscopic radioautograph of a section of cultured rat aortic smooth muscle cells which had been incubated for 48 h with VLDL remnant particles labeled with ^{125}I. The radioautographic reaction is concentrated over a secondary lysosome (arrow). × 17,000. From *Bierman et al.* [28].

Catabolism of Serum Lipoproteins

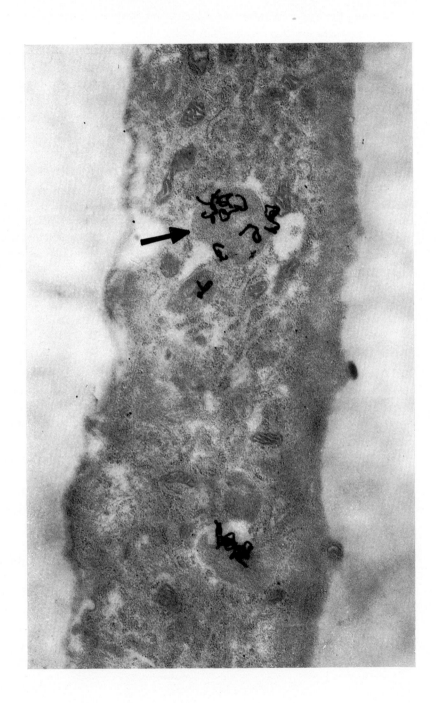

Among the extrahepatic tissues which come in direct contact with the remnant particles are the various endothelia lining the blood vessels. The endothelial cells should be able to deal with remnant particles in two ways: the particle could be transported from the luminal side to the abluminal side of endothelium by vesicular transport or it could be interiorized by the endothelial cell proper and undergo intralysosomal degradation. In the first instance, the remnant particles would come in contact with extravascular elements, and in the case of the artery, also with other cellular elements of the vessel wall. Studies with cultured rat aortic smooth muscle cells have shown indeed that these cells take up VLDL remnants much more extensively than intact VLDL [27]. Following interiorization, the remnant particles were localized by radioautography to secondary lysosomes of the smooth muscle cell (fig. 2), while labeled degradation products were recovered in the culture medium. The ability of endothelial cells to interact with VLDL remnants was investigated in cultured endothelial cells derived from human umbilical veins. The interaction resulted in intracellular uptake and in some instances in the enrichment of the cells in cholesterol ester [28].

Low-Density Lipoprotein Catabolism

Normal Pathway

In the rat, most of the remnant cholesterol ester is removed by the liver [19, 20] and its catabolism regulates hepatic sterol synthesis [29]. In man, however, further catabolism of the VLDL remnant occurs at an extracellular site and results in the formation of low-density lipoproteins (LDL) [30]. The role of LDL as a cholesterol donor *in vivo* has been accepted in general, but the elucidation of the role of LDL catabolism in this process has come only from studies in tissue culture, which have culminated in the evolution of the concept of receptor-mediated control of LDL metabolism [31]. Investigations of the regulation of cholesterol synthesis in cultured cells provided evidence that exogenous free cholesterol can modulate intracellular synthesis of cholesterol from acetate, but not from mevanolate and hence focused the attention to 3-hydroxy-3-methyl-glutaryl coenzyme A (HMG CoA) reductase [32–34]. It became apparent that LDL as well can serve as a modulator of HMG CoA reductase levels in cultured cells [35] and these findings prompted further elucidation of the pathway by which cholesterol is transferred from the lipoprotein to the site at which it regulates intracellular sterol synthesis. This then was the background for the development of the by now well known pathway in which LDL participates in this process. The pathway consists of three steps: surface binding to a specific receptor, followed by endocytosis and intracellular degradation. The last step involves hydrolysis of the apoproteins to amino acids [36] as well as hydrolysis of cholesterol ester and formation of free cholesterol [37]. This last reaction is apparently the key step in the process of regulation not only of sterol synthesis, but also of new receptor synthesis [31]. Thus a built-in mechanism of feedback regulation of LDL catabolism was shown to operate in cultured cells. At the cellular level, the site of hydrolysis of the LDL particles as well as other lipoproteins which are endocytosed by cultured smooth muscle cells or fibroblasts, was shown to be the lysosome [27, 38, 39]. In more recent studies, the site of the LDL receptor has been localized to coated vesicles abutting on the plasma membrane [40].

Genetic Defects

As in many other instances, validation of the various steps described in the normal pathway came from the study of genetic aberrations. Thus, it became evident that the apparent impairment of LDL degradation in homozygous familial hypercholesteremia was not due to a lysosomal defect, but to the failure to deliver the lipoprotein to the site of degradation, i.e. the lysosome [36]. The defects which are responsible for this impairment were related to a lack, a reduced number or abnormal localization of surface receptors which result in reduction of binding and interiorization or interiorization alone [41]. LDL catabolism was shown to be impaired in fibroblasts derived from a patient with Wolman's disease in which the main defect is a reduction of lysosomal acid lipase and cholesterol ester hydrolase [42].

Changes in the LDL Particle

Since the receptor-mediated pathway of LDL catabolism depended extensively on the recognition of the LDL particle, attempts were made to elucidate this process by modulation of the protein moiety of the particle. Thus, acetylation of LDL which increased the net negative charge of the particle eliminated its ability to bind to cellular receptors and its subsequent interiorization [43]. On the other hand, cationization of LDL which adds positive charges to the particle resulted in enhancement of LDL binding [43]. The positively charged LDL particles are taken up by a pathway not dependent on the existence of the specific LDL receptor, as cells from normal and HFH donors interiorize and degrade cationized LDL to the same extent [43].

These studies with the cationized LDL have indicated that LDL catabolism may not be dependent solely on the receptor-mediated pathway. Thus a change in the surface charge or some other perturbation can lead to enhanced LDL interiorization by alternative pathways and subsequent lysosomal degradation. Recently, partial tryptic digestion of LDL was shown to promote its degradation by HFH cells without altering the binding of the particle to the surface of cells lacking the receptor [44]. It was proposed that similar alterations could precede the nonreceptor-mediated uptake and catabolism of LDL *in vivo*.

Another type of alteration in the LDL particle was induced by cholesterol feeding to rabbits. These LDL particles promoted growth of aortic cells in primary culture [45]. It is not clear whether the growth-promoting property of hyperlipidemic LDL is related to its catabolism or to the presence in the LDL density range of arginine-rich peptide [45]. It seems relevant to close this section on 'Changes in the LDL Particle' with some remarks on the fact that the lipoproteins isolated in the LDL density range may not be quite homogeneous also under normal conditions. Recent studies have shown the presence of immunoregulatory LDL with a solvent density of 1.055 g/ml which do profoundly affect certain surface properties of lymphocytes [46, 47].

LDL Catabolism and Atherosclerosis

The hallmark of the atherosclerotic lesion is the accumulation of intracellular and extracellular cholesterol ester. The accumulation of cholesterol ester has been linked to an increased ingestion of lipoproteins, but the subsequent steps have not been elucidated completely. Cellular accumulation of cholesterol ester could result from the reesterification of the free cholesterol, a product of normal intralysosomal hydrolysis of cholesterol ester. Alternately, an impairment in the degradation of LDL cholesterol ester could also contribute towards the intracellular accumulation of cholesterol ester [48]. In the first instance, the accumulated cholesterol ester would be localized in the cytoplasmic compartment, while in the second, both intralysosomal and extralysosomal accumulation might be expected. In the human atheroma, both forms of accumulation have been encountered, but the quantitation of the contribution of the intra- versus extralysosomal form towards the overall cholesterol ester accumulation on a purely morphological basis is rather difficult. Therefore, conclusions were drawn from determinations of fatty acid composition of the accumulating cholesterol ester. Since in the LDL, cholesterol linoleate is the main cholesterol ester, while in the fatty streak cholesterol oleate predominates [49], these findings were taken to support the first hypothesis, i.e. that accumulation results only from extralysosomal reesterification. However, the primary defect could lie in increased influx and relative

Table I. Retention and degradation of ^{125}I-labeled LDL by chloroquine-treated human skin fibroblasts.

Additions to medium		^{125}I-LDL protein, ng/mg cell protein per 48 h	
LDL protein µg/ml	chloroquine µM	retention	degradation
10	0	73	1,700
10	70	1,733	250
360	0	2,000	12,300
360	70	26,600	4,633

The cells were grown in medium containing 10% fetal calf serum throughout the experimental period. When the cells reached confluency, chloroquine and LDL were added and incubations were continued for 48 h. Values are means of duplicate Petri dishes. Retention $=^{125}$I-protein recovered from the cells. For other details, see *Stein et al.* [51, adapted from table I].

impairment of lysosomal LDL cholesterol ester hydrolysis, which, in the beginning, might lead to predominantly intralysosomal accumulation. With time, the influx might decrease, and so relatively more of the intralysosomal cholesterol ester would undergo hydrolysis and reesterification in the cytoplasm. If at the same time cholesterol efflux would fall even more than the influx, this would contribute further towards the accumulation of cholesterol oleate. Another explanation for the accumulation of cholesterol oleate rather than linoleate was based on findings that in a cell-free system highly unsaturated cholesterol esters are less resistant to hydrolysis than relatively saturated esters [50]. However, recent studies in cultured human skin fibroblasts have shown that this does not occur *in situ,* as will be discussed in the next section.

In order to study the possible role of impaired LDL catabolism in the formation of atherosclerotic lesions we have exposed cells in culture to LDL as cholesterol donor and chloroquine [51, 52], an inhibitor of lysosomal hydrolases [53]. As seen in Table I, addition of chloroquine to the culture medium resulted in an inhibition of degradation of LDL apoprotein and in the presence of a high concentration of LDL, in an accumulation of labeled intracellular protein. Under such experimental conditions there was also accumulation of cholesterol and cholesterol ester and this proved to be quite reproducible in cells derived from different donors (table II). At the ultrastructural level, accumulation of material in membrane-bound vacuoles and an increase in

Table II. Effect of LDL and chloroquine on cholesterol accumulation in human skin fibroblasts

Cell donor	Passage No.	Protein/dish µg	Cholesterol µg/mg cell protein	
			free	ester
M	2	340	68.5	85.2
	4	418	84.8	85.2
	5	374	72.9	99.6
D	2	520	57.0	83.2
	3	360	59.1	89.3
A	4	420	79.1	101.3
	8	395	86.0	90.6

Conditions of culture as in table I. During the 48 h of incubation the cells were exposed to 50 µM chloroquine and 400 µg LDL protein/ml. Values are means of triplicate determinations.

cytoplasmic lipid droplets was found. Similar lesions and accretion of cholesterol ester were found also in aortic smooth muscle cells subjected to LDL and chloroquine (fig. 3). What relation do these inclusions bear to lesions described in human or experimental atherosclerosis? There seems to be a more general agreement that the foam cells of atherosclerotic lesions are derived from smooth muscle cells [54]. Studies have shown that a certain portion of the foam cell 'vacuoles' do contain acid phosphatase [48] and other acid hydrolases [55] and thus can be classified as secondary lysosomes. The morphological finding in the cultured aortic smooth muscle cells treated with LDL and chloroquine [52] and cells in experimental atheromatosis in rabbits [48] or monkeys [56, 57] show certain similarities with respect to the electron-dense intracellular inclusions, which had been defined as lysosomes. Similar structures are seen also in electron micrographs [58, 59] which represent sections of human fatty streaks. Thus it seems that the chloroquine-treated

Fig. 3. Electron micrograph of cultured rat aortic smooth muscle cells exposed to 50 µM chloroquine and 350 µg protein per milliliter of LDL for 48 h. The cell on the right is distended with inclusions of various electron density, which correspond to the finding of accumulated cholesterol ester. × 26,000.

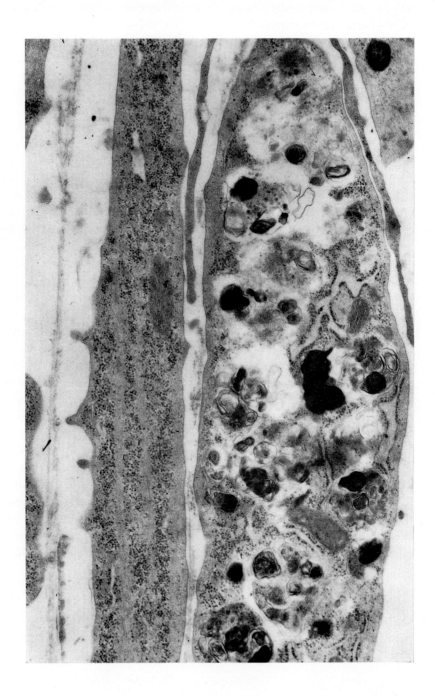

cells provide a valid model to study the thesis whether a relative insufficiency of lysosomal enzymes coupled with increased influx of LDL could be the basic pathogenic mechanism of atheromatosis. Since both the LDL load and chloroquine are easily removed from the culture medium, it became possible to study some of the requirements for the removal of cholesterol from the cells [51, 52].

The above-described experimental design was applied also to test the thesis whether cholesterol linoleate is hydrolyzed preferentially to other cholesterol esters. The cultured human skin fibroblasts were incubated with chloroquine and high concentrations of LDL for 48 h and allowed to accumulate considerable amounts of cholesterol ester. They were then postincubated in the presence of cholesterol acceptors in the medium, to promote intracellular hydrolysis. At the end of the accumulation period, the ratio between cholesterol linoleate and oleate was 3.5 or 3.1 as determined by the fatty acid composition of cellular cholesterol ester. When the cholesterol ester content had decreased by 43–47%, this ratio fell to 2.3 or 1.4. Such a finding could have indicated a preferential hydrolysis of cholesterol linoleate or a comparable hydrolysis, of both esters followed by a preferential reesterification of the free cholesterol with oleic acid. To test these possibilities, LDL was labeled with ^{3}H-cholesterol esters of varying fatty acid composition (linoleate, oleate, palmitate and stearate) and used as cholesterol ester donor during exposure to chloroquine. Practically all the label was found in the form of cholesterol ester at the end of 24 h incubation with chloroquine. During the postincubation period, in the absence of chloroquine in the medium, hydrolysis of the labeled cholesterol esters occurred as shown by the formation of labeled free cholesterol. Comparison of the hydrolysis of the four labeled cholesterol esters indicated that under these experimental conditions all four cholesterol esters were hydrolyzed at the same rate. Thus, it seems that the decrease in the cholesterol linoleate:cholesterol oleate ratio after removal of chloroquine was not due to preferential hydrolysis of cholesterol linoleate by the lysosomal cholesterol esterase, but rather to a reesterification of the free cholesterol with oleic acid. In analogy, in organ cultures of atherosclerotic rabbit aorta a similar disappearance rate of cholesterol linoleate and oleate was found during a chase of 10 days [60].

LDL catabolism in vivo

Even though it is accepted that an LDL particle is derived from a VLDL particle, the site and exact mode of LDL formation still remain to be elucidated. The site of LDL degradation in the intact organism was studied in pigs

[61] and rats [20], with the emphasis on the delineation of the role of the liver in this process. When LDL was injected into pigs, it was cleared from the circulation with a $t_{1/2}$ of 19.3 h [61]. As hepatectomy did not prolong the clearance of LDL from the plasma, it was concluded that the liver does not play an important role in LDL catabolism, which apparently occurs mainly in extrahepatic tissues [61]. In the rat, we have used chloroquine, the pretreatment with which inhibited degradation of VLDL remnant particles [20], in order to evaluate the participation of the liver in LDL degradation. When 50% of the injected labeled LDL had been cleared from the circulation, not more than 4.4% were recovered in the liver and this amount was increased to not more than 10% in chloroquine-treated rats. More recent studies *in vivo* were directed towards the question in which organs of the intact rat LDL participates in the regulation of sterol synthesis. The rats were pretreated with an adenine analog 4-aminopyrazole-[3,4d]pyrimidine (APP), which is known to reduce plasma cholesterol lipoprotein levels within 48 h after administration [62]. This reduction was accompanied by a marked stimulation in the activity of HMG CoA reductase in kidneys and lungs and an elevation in the rate of cholesterol synthesis [63]. In another study following reduction of serum lipoprotein levels by APP pretreatment, sterol synthesis from acetate was studied in various extrahepatic tissues before and after infusion of homologous lipoproteins of d < 1.23 g/ml or human LDL [64]. The results obtained indicate that LDL regulates sterol synthesis in the rat intestine, kidney, lung, adrenal and ovary. Thus, if regulation is achieved through the same process of internalization and intracellular degradation of the LDL particle as shown for cultured fibroblasts, then these organs as well as other peripheral tissues should be considered as the site of LDL catabolism.

Catabolism of HDL

HDL is the major lipoprotein class in the rat and its catabolism in this species was studied both *in vivo* and in tissue culture. Following injection of ^{125}I-HDL, the disappearance curve showed at least two components. Screening by passage through a rat for 12 h eliminated the fast component and so the $t_{1/2}$ of the HDL protein in the circulation was found to be 10.50 h [65]. Attempts were made to determine whether the fast component, observed in the disappearance of unscreened HDL, might be due to a preferential loss of one or more apolipoproteins. However, the lack of change in the distribution of label among the main apolipoproteins at different times after injection, supports the thesis that the HDL is cleared from the circulation as an intact particle [66]. About 10% of the injected HDL was recovered in the liver and about 5% in the small intestine 6 h after injection [65]. Data from isolated perfused rat liver suggest that about 7% of apoHDL is catabolized by the liver [67]. Following intravenous injection into intact rats, most of the label present in the liver was in the form of protein-bound radioactivity. With the help of radioautography, the label was found predominantly over parenchymal liver cells and some label was seen also over endothelial and Kupffer cells [68]. The intracellular localization of the labeled HDL was studied by electron-microscopic radioautography. It was postulated that HDL was most probably taken up by the process of pinocytosis and transported towards the site of its catabolism which was localized to secondary lysosomes [68]. Similar localization was also seen in the epithelial cells of the small intestine. The role of hepatic lysosomal enzymes in the degradation of HDL apoproteins was confirmed also by studies *in vitro* using a sonicated postnuclear supernatant of rat liver [20]. Human and rat HDL as well as their delipidated forms were hydrolyzed very actively at acid pH. The degradation could be inhibited 50-90% by addition of 50-100 mM chloroquine to the incubation medium. These results indicate that cathepsin B_1 is involved in the catabolism of HDL [20]. In the dog liver lysosomes, cathepsin B and D were shown to degrade HDL_3 [69].

As in the rat the liver could account for only a fraction of HDL catabolism its degradation by extrahepatic tissues was also determined using cells in culture. Exposure of rat aortic smooth muscle cells to homologous HDL resulted in interiorization of the labeled lipoprotein and its degradation [38, 39]. In comparison to homologous LDL, less HDL was metabolized during a comparable period of time [70]. These studies have also shown that even though human LDL was bound and taken up by the rat cells, little or no internalization of human HDL had occurred [70]. Similar findings were reported also for swine aortic smooth muscle cells exposed to homologous lipoproteins [71]. Degradation of human HDL was studied quite extensively using human skin fibroblasts in culture. In contradistinction to human LDL, HDL was interiorized mainly by a process of fluid endocytosis, which may not require a specific receptor [72]. Why then was rat HDL bound and taken up more extensively than human HDL by human cells as shown in another comparative study [73]. This discrepancy became elucidated following an elegant series of studies which had led to the discovery that the high affinity LDL receptor can also recognize the ARP present in other lipoproteins [74]. In those studies pathological lipoproteins were produced by high cholesterol diets that were fed to pigs, dogs, rats or monkeys [75–78]. These had a mobility on electrophoresis and were present in the low-density range, but were devoid of apoB protein. These lipoproteins, designated HDLc, were found to contain high amounts of ARP and were shown to compete for the LDL receptor [79]. The HDLc was found to be actively degraded by peripheral cells [74]. As normal rat HDL contains ARP in contradistinction to human HDL, it would be taken up by the LDL receptors. Indeed a fraction isolated from delipidated rat HDL which was highly enriched in ARP (HS-2), was taken up more avidly by rat aortic smooth muscle cells than whole delipidated rat or human HDL [39].

One should consider also the possibility of dissociation of HDL apoproteins from the parent particle as a step in the degradation process. The subsequent catabolism of the apoproteins might proceed independently and indeed apoA-I and apoA-II did show different affinity for binding to cell surfaces than the apoC apoproteins [80]. In an attempt to extrapolate the findings in culture to conditions *in vivo,* an estimate was made of a potential degradation of HDL by 'peripheral cells' and it was calculated to exceed the estimated turnover rate for HDL *in vivo* [72]. However, one has to bear in mind the fact that the activity of lysosomal enzymes in cultured cells, when related to cell mass, exceeds by far that found in their *in vivo* counterparts [81]. If one tries to relate the catabolism of HDL to a functional parameter as discussed for LDL with respect to regulation of sterol synthesis, then it becomes apparent that in most tissues, human HDL does not play an important role in this respect. In the rat, the two exceptions are the ovary and adrenal in which both human and rat HDL is apparently the principal lipoprotein which regulates sterol synthesis [64]. There might be, however, species differences as HDL had no effect in adrenal cells derived from mice [82]. The exact site of HDL degradation in the intact organism has not been delineated so far and one should consider also the participation of the reticuloendothelial system in this process. Indeed, following perfusion of rat hearts with labeled HDL, concentrations of label could be shown over secondary lysosomes of interstitial macrophages [83].

Acknowledgement

This investigation was supported in part by a grant from the Israeli Ministry of Health and from the Délégation Générale à la Recherche Scientifique et Technique of the French Government.

O.S. and Y.S. are established investigators of the Israeli Ministry of Health.

References

1 Robinson, D. S.: Removal of triglyceride fatty acids from the blood; in Comprehensive biochemistry, vol. XVII, pp. 51–116 (Elsevier, Amsterdam 1970).
2 Zinder, O.; Mendelson, C. R.; Blanchette-Mackie, E. J., and Scow, R. O.: Lipoprotein lipase and uptake of chylomicron triacylglycerol and cholesterol by perfused rat mammary tissue. Biochim. biophys. Acta *431:* 526–537 (1976).
3 Windmueller, H. G.; Herbert, P. N., and Levy, R. I.: Biosynthesis of lymph and plasma lipoprotein apoproteins by isolated perfused rat liver and intestine. J. Lipid. Res. *14:* 215–223 (1973).
4 Imaizumi, K.; Fainaru, M., and Havel, R. J.: Composition of proteins of mesenteric lymph chylomicrons in the rat and alterations produced upon exposure of chylomicrons to blood serum and serum lipoproteins. J. Lipid Res. *19:* 712–722 (1978).
5 Schoefl, G. I. and French, J. E.: Vascular permeability to particulate fat: morphological observations on vessels of lactating mammary gland and of lung. Proc. R. Soc. B *169:* 153–163 (1968).
6 Blanchette-Mackie, E. J. and Scow, R. O.: Sites of lipoprotein lipase activity in adipose tissue perfused with chylomicrons. J. Cell Biol. *51:* 1–25 (1971).
7 Chajek, T.; Stein, O., and Stein, Y.: Interaction of concanavalin A with membrane bound and solubilized lipoprotein lipase of rat heart. Biochim. biophys. Acta *431:* 507–518 (1976).
8 Schotz, M. C.; Twu, J. S.; Pedersen, M. E.; Chen, C. H.; Garfinkel, A. S., and Borensztajn, J.: Antibodies to lipoprotein lipase application to perfused heart. Biochim. biophys. Acta *489:* 214–224 (1977).
9 Chajek, T.; Stein, O., and Stein, Y.: Interference with the transport of heparin releasable lipoprotein lipase in the perfused rat heart by colchicine and vinblastine. Biochim. biophys. Acta *388:* 260–267 (1975).
10 Cryer, A.; McDonald, A.; Williams, E. R., and Robinson, D. S.: Colchicine inhibition of the heparin-stimulated release of clearing-factor lipase from isolated fat-cells. Biochem. J. *152:* 717–720 (1975).
11 Chajek, T.; Stein, O., and Stein, Y.: Rat heart in culture as a tool to elucidate the cellular origin of lipoprotein lipase. Biochim. biophys. Acta *488:* 140–144 (1977).
12 Chajek, T.; Stein, O., and Stein, Y.: Lipoprotein lipase of cultured mesenchymal rat heart cells. I. Synthesis, secretion and releasability by heparin. Biochim. biophys. Acta *528:* 456–465 (1978).
13 Chajek, T.; Stein, O., and Stein, Y.: Lipoprotein lipase of cultured mesenchymal rat heart cells. II. Hydrolysis of labeled very low density lipoprotein triacylglycerol by membrane supported enzyme. Biochim. biophys. Acta *528:* 466–474 (1978).
14 Fielding, C. J. and Higgins, J. M.: Lipoprotein lipase; comparative properties of the membrane-supported and solubilized enzyme species. Biochemistry, N. Y. *13:* 4324–4329 (1974).
15 Vogel, W. C. and Zieve, L.: Postheparin phospholipase. J. Lipid Res. *5:* 177–183 (1964).
16 Redgrave, T.: Formation of cholesteryl ester-rich particulate lipid during metabolism of chylomicrons. J. clin. Invest. *49:* 465–471 (1970).
17 Eisenberg, S.; Bilheimer, D. W.; Lindgren, F. T., and Levy, R. I.: On the metabolic

conversion of human plasma very low density lipoprotein to low density lipoprotein. Biochim. biophys. Acta *326:* 361–377 (1973).
18 Eisenberg, S. and Rachmilewitz, D.: Interaction of rat plasma very low density lipoprotein with lipoprotein lipase-rich (postheparin) plasma. J. Lipid Res. *16:* 341–351 (1975).
19 Quarfordt, S.H. and Goodman, D.S.: Metabolism of doubly-labeled chylomicron cholesteryl esters in the rat. J. Lipid Res. *8:* 264–272 (1967).
20 Stein, Y.; Ebin, V.; Bar-On, H., and Stein, O.: Chloroquine induced interference with degradation of serum lipoproteins in rat liver, studied *in vivo* and *in vitro*. Biochim. biophys. Acta *486:* 286–297 (1977).
21 Eisenberg, S. and Rachmilewitz, D.: Metabolism of rat plasma very low density lipoprotein. I. Fate in circulation of the whole lipoprotein. Biochim. biophys. Acta *326:* 378–390 (1973).
22 Stein, O.; Stein, Y.; Fidge, A., and Goodman, D.S.: The metabolism of chylomicron cholesteryl ester in rat liver. A combined radioautographic-electron microscopic and biochemical study. J. Cell Biol. *43:* 410–431 (1969).
23 Stein, O.; Rachmilewitz, D.; Sanger, L.; Eisenberg, S., and Stein, Y.: Metabolism of iodinated very low density lipoprotein in the rat; autoradiographic localization in the liver. Biochim. biophys. Acta *360:* 205–216 (1974).
24 Nilsson, A.: Effects of anti-microtubular agents and cycloheximide on the metabolism of chylomicron cholesteryl esters by hepatocyte suspensions. Biochem. J. *162:* 367–377 (1977).
25 Pesonti, E.C. and Axline, S.G.: Phagolysosome formation in normal and colchicine treated macrophages. J. exp. Med. *142:* 903–913 (1975).
26 Chajek, T.; Friedman, G.; Stein, O., and Stein, Y.: Effect of colchicine, cycloheximide and chloroquine on the hepatic triacylglycerol hydrolase in the intact rat and perfused liver. Biochim. biophys. Acta *488:* 270–279 (1977).
27 Bierman, E.L.; Eisenberg, S.; Stein, O., and Stein, Y.: Very low density lipoprotein 'remnant' particles: uptake by aortic smooth muscle cells in culture. Biochim. biophys. Acta *329:* 163–169 (1973).
28 Stein, Y.; Stein, O., and Goren, R.: Metabolism and metabolic role of serum high density lipoproteins; in Gotto, Miller and Oliver, High density lipoproteins and atherosclerosis, pp. 37–49 (North-Holland Biomedical Press, Amsterdam 1978).
29 Andersen, J.M. and Dietschy, J.M.: Regulation of sterol synthesis in 16 tissues of rat. I. Effect of diurnal light cycling, fasting, stress, manipulation of enterohepatic circulation, and administration of chylomicrons and triton. J. biol. Chem. *252:* 3646–3651 (1977).
30 Eisenberg, S.: Lipoprotein metabolism and hyperlipemia; in Paoletti and Gotto, Atherosclerosis reviews, vol. 1, pp. 23–61 (Raven Press, New York 1976).
31 Brown, M.S. and Goldstein, J.L.: Receptor-mediated control of cholesterol metabolism. Study of human mutants has disclosed how cells regulate a substance that is both vital and lethal. Science *191:* 150–191 (1976).
32 Bailey, J.M.: Lipid metabolism in cultured cells. VI. Lipid biosynthesis in serum and synthetic growth media. Biochim. biophys. Acta *125:* 226–236 (1966).
33 Rothblat, G.H.: The effect of serum components on sterol biosynthesis in L cells. J. cell. Physiol. *74:* 163–170 (1969).
34 Williams, C.D. and Avigan, J.: *In vitro* effects of serum proteins and lipids on lipid

synthesis in human skin fibroblasts and leukocytes grown in culture. Biochim. biophys. Acta *260:* 413–423 (1972).

35   Goldstein, J.L. and Brown, M.S.: Familial hypercholesterolemia: identification of a defect in the regulation of 3-hydroxy-3-methylglutaryl coenzyme A reductase activity associated with overproduction of cholesterol. Proc. natn. Acad. Sci. USA *70:* 2804–2808 (1973).

36   Goldstein, J.L. and Brown, M.S.: Binding and degradation of low density lipoproteins by cultured human fibroblasts. Comparison of cells from a normal subject and from a patient with homozygous familial hypercholesterolemia. J. biol. Chem. *249:* 5153–5162 (1974).

37   Brown, M.S.; Dana, S.E., and Goldstein, J.L.: Receptor-dependent hydrolysis of cholesteryl esters contained in plasma low density lipoprotein. Proc. natn. Acad. Sci. USA *72:* 2925–2929 (1975).

38   Bierman, E.L.; Stein, O., and Stein, Y.: Lipoprotein uptake and metabolism by rat aortic smooth muscle cells in tissue culture. Circulation Res. *35:* 136–150 (1974).

39   Stein, O. and Stein, Y.: Comparative uptake of rat and human serum low density lipoproteins by rat aortic smooth muscle cells in culture. Circulation Res. *36:* 436–443 (1975).

40   Anderson, R.G.W.; Brown, M.S., and Goldstein, J.L.: Role of the coated endocytic vesicle in the uptake of receptor-bound low density lipoprotein in human fibroblasts. Cell *10:* 351–364 (1977).

41   Brown, M.S.; Anderson, R.G.W., and Goldstein, J.L.: Mutations affecting the binding, internalization, and lysosomal hydrolysis of low density lipoprotein in cultured human fibroblasts, lymphocytes, and aortic smooth muscle cells. J. supramolec. Struct. *6:* 85–94 (1977).

42   Goldstein, J.L.; Dana, S.E.; Faust, J.R.; Beaudet, A.L., and Brown, M.S.: Role of lysosomal acid lipase in the metabolism of plasma low density lipoprotein. Observations in cultured fibroblasts from a patient with cholesteryl ester storage disease. J. biol. Chem. *250:* 8487–8495 (1975).

43   Basu, S.K.; Goldstein, J.L.; Anderson, R.G.W., and Brown, M.S.: Degradation of cationized low-density lipoprotein and regulation of cholesterol metabolism in homozygous familial hypercholesterolemia fibroblasts. Proc. natn. Acad. Sci. USA *73:* 3178–3182 (1976).

44   Carew, T.E.; Chapman, J., and Steinberg, D.: Enhanced degradation of proteolytically damaged low density lipoprotein (LDL) by fibroblasts from patients with homozygous familial hypercholetesrolemia (HFH). Abstr. Circulation, No. 380, pp. IIII–99 (1977).

45   Chen, R.M.; Getz, G.S.; Fischer-Dzoga, K., and Wissler, R.W.: The role of hyperlipidemic serum on the proliferation and necrosis of aortic medial cells *in vitro.* Expl molec. Path. *26:* 359–374 (1977).

46   Curtiss, L.A. and Edgington, T.E.: Regulatory serum lipoproteins: regulation of lymphocyte stimulation by a species of low density lipoprotein. J. Immun. *116:* 1452–1458 (1976).

47   Curtiss, L.K.; DeHeer, D.H., and Edgington, T.S.: *In vivo* suppression of the primary immune response by a species of low density serum lipoprotein. J. Immun. *118:* 648–652 (1977).

48 Shio, H.; Farquhar, M.G., and De Duve, C.: Lysosomes of the arterial wall. IV. Cytochemical localization of acid phosphatase and catalase in smooth muscle cells and foam cells from rabbit atheromatous aorta. Am. J. Path. *76:* 1–16 (1974).

49 Smith, E.B.: The influence of age and atherosclerosis on the chemistry of aortic intima. J. Atheroscler. Res. *5:* 224–240 (1965).

50 Takano, T.; Black, W.J.; Peters, T.J., and De Duve, C.: Assay, kinetics and lysosomal localization of an acid cholesteryl esterase in rabbit aortic smooth muscle cells. J. biol. Chem. *249:* 6732–6737 (1974).

51 Stein, O.; Vanderhoek, J.; Friedman, G., and Stein, Y.: Deposition and mobilization of cholesterol ester in cultured human skin fibroblasts. Biochim. biophys. Acta *450:* 367–378 (1976).

52 Stein, O.; Vanderhoek, J., and Stein, Y.: Cholesterol ester accumulation in cultured aortic smooth muscle cells. Induction of cholesterol ester retention by chloroquine and low density lipoprotein and its reversion by mixtures of high density apolipoprotein and sphingomyelin. Atherosclerosis *26:* 465–482 (1977).

53 Wibo, M. and Poole, B.: Protein degradation in cultured cells. II. The uptake of chloroquine by rat fibroblasts and the inhibition of cellular protein degradation and cathepsin B_1. J. Cell Biol. *63:* 430–440 (1974).

54 Wissler, R.W.: The arterial medial cell, smooth muscle or multifunctional mesenchyme? J. Atheroscler. Res. *8:* 201–213 (1968).

55 Peters, T.J. and De Duve, C.: Lysosomes of the arterial wall. II. Subcellular fractionation of aortic cells isolated from rabbits with experimental atheroma. Expl. molec. Path. *20:* 228–256 (1974).

56 Tucker, C.F.; Catsulis, C.; Strong, J.P., and Eggen, D.A.: Regression of early cholesterol-induced aortic lesions in rhesus monkeys. Am. J. Path. *65:* 493–502 (1971).

57 Goldfischer, S.; Schiller, B., and Wolinsky, H.: Lipid accumulation in smooth muscle cell lysosomes in primate atherosclerosis. Am. J. Path. *78:* 497–502 (1975).

58 Geer, J.C.; McGill, H.R., and Strong, J.P.: The fine structure of human atherosclerotic lesions. Am. J. Path. *38:* 263–275 (1961).

59 Geer, J.C. and Haust, M.D.: Smooth muscle cells in atherosclerosis. Monogr. Atheroscler., vol. 2 (Karger, Basel 1972).

60 Hudson, K.; Day, A.J., and Horsch, A.K.: Removal of fatty acid labelled cholesterol ester, phospholipid and triglyceride from atherosclerotic rabbit aorta *in vitro*. Atherosclerosis *28:* 425–434 (1977).

61 Sniderman, A.D.; Carew, T.E.; Chandler, J.G., and Steinberg, D.: Paradoxical increase in rate of catabolism of low-density lipoproteins after hepatectomy. Science *183:* 526–528 (1974).

62 Shiff, T.S.; Roheim, P.S., and Eder, H.A.: Effects of high sucrose diets and 4-aminopyrazolopyrimidine on serum lipids and lipoprotein in the rat. J. Lipid. Res *12:* 596–603 (1971).

63 Balasubramaniam, S.; Goldstein, J.L.; Faust, J.R., and Brown, M.S.: Evidence for regulation of 3-hydroxy-3-methylglutaryl coenzyme A reductase activity and cholesterol synthesis in nonhepatic tissues of rat. Proc. natn. Acad. Sci. USA *73:* 2564–2568 (1976).

64 Andersen, J.M. and Dietschy, J.M.: Regulation of sterol synthesis in 15 tissues of rat. II. Role of rat and human high and low density plasma lipoproteins and of rat chylomicron remnants. J. biol. Chem. *252:* 3652–3659 (1977).

65 Roheim, P.S.; Rachmilewitz, D.; Stein, O., and Stein, Y.: Metabolism of iodinated high density lipoproteins in the rat. I. Half-life in the circulation and uptake by organs. Biochim. biophys. Acta 248: 315–329 (1971).

66 Roheim, P.S.; Hirsch, H.; Edelstein, D., and Rachmilewitz, D.: Metabolism of iodinated high density lipoprotein subunits in the rat. III. Comparison of the removal rate of different subunits from the circulation. Biochim. biophys. Acta 278: 517–529 (1972).

67 Sigurdsson, G.; Noel, S.-P., and Havel, R.J.: Quantification of the hepatic contribution to the catabolism of high density apolipoprotein (Apo-HDL) in the rat. Abstract. Circulation 55/56: suppl. III, No. 3, III4 (1977).

68 Rachmilewitz, D.; Stein, O.; Roheim, P.S., and Stein, Y.: Metabolism of iodinated high density lipoproteins in the rat. II. Autoradiographic localization in the liver. Biochim. biophys. Acta 270: 414–425 (1972).

69 Nakai, T.; Otto, P.S., and Whayne, T.F., jr.: Proteolysis of canine apolipoprotein by acid proteases in canine liver lysosomes, Biochim. biophys. Acta 422: 380–389 (1976).

70 Stein, O. and Stein, Y.: Surface binding and interiorization of homologous and heterologous serum lipoproteins by rat aortic smooth muscle cells in culture. Biochim. biophys. Acta 398: 377–384 (1975).

71 Carew, T.E.; Koschinsky, T.; Hayes, S.B., and Steinberg, D.: A mechanism by which high-density lipoproteins may slow the atherogenic process. Lancet i: 1315–1317 (1976).

72 Miller, N.E.; Weinstein, D.B., and Steinberg, D.: Binding, internalization, and degradation of high density lipoprotein by cultured normal human fibroblasts. J. Lipid Res. 18: 438–450 (1977).

73 Stein, Y.; Stein, O., and Vanderhoek, J.: Role of serum lipoproteins in the transport of cellular cholesterol; in Greten, Lipoprotein metabolism, pp. 99–105 (Springer, Berlin 1976).

74 Bersot, T.P.; Mahley, R.W.; Brown, M.S., and Goldstein, J.L.: Interaction of swine lipoproteins with the low density lipoprotein receptor in human fibroblasts. J. biol. Chem. 251: 2395–2398 (1976).

75 Mahley, R.W.; Weisgraber, K.H., and Innerarity, T.L.: Canine lipoproteins and atherosclerosis. II. Characterization of the plasma lipoproteins associated with atherogenic and non-atherogenic hyperlipidemia. Circulation Res. 35: 722–733 (1974).

76 Mahley, R.W.; Weisgraber, K.H., and Innerarity, T.L.: Atherogenic hyperlipoproteinemia induced by cholesterol feeding in the patas monkey. Biochemistry, N.Y. 15: 2979–2985 (1976).

77 Mahley, R.W.; Weisgraber, K.H.; Innerarity, T.L.; Brewer, H.B., jr., and Assman, G.: Swine lipoproteins and atherosclerosis. Changes in the plasma lipoproteins and apoproteins induced by cholesterol feeding. Biochemistry, N.Y. 14: 2817–2823 (1975).

78 Mahley, R.W. and Holcombe, K.S.: Alterations of plasma lipoproteins and apoproteins following cholesterol feeding in the rat. J. Lipid Res. 18: 314–324 (1977).

79 Mahley, R.W. and Innerarity, T.L.: Interaction of canine and swine lipoproteins with the low density lipoprotein receptor of fibroblasts as correlated with heparin/manganese precipitability. J. biol. Chem. 252: 3980–3986 (1977).

80 Jackson, R.L.; Gotto, A.M.; Stein, O., and Stein, Y.: A comparative study on the

removal of cellular lipids from Landschutz ascites cells by human plasma apolipoproteins. J. biol. Chem. *250:* 7204–7209 (1975).

81 Fowler, S.; Shio, H., and Wolinsky, H.: Subcellular fractionation and morphology of calf aortic smooth muscle cells. Studies on whole aorta, aortic explants, and subcultures grown under different conditions, J. Cell Biol. *75:* 166–184 (1977).

82 Faust, J.R.; Goldstein, J.L., and Brown, M.S.: Receptor-mediated uptake of low density lipoprotein and utilization of its cholesterol for steroid synthesis in cultured mouse adrenal cells. J. biol. Chem. *252:* 4861–4871 (1977).

83 Stein, O. and Stein, Y.: Synthesis and intracellular degradation of serum lipoproteins: in Handbook of internal medicine, vol. 7., pp. 197–218 (Springer, Berlin 1976).

O. Stein, MD, Lipid Research Laboratory, Department of Medicine B, Hadassah University Hospital, Jerusalem (Israel)

Effect of Hypolipidemic Drugs on Serum Lipoproteins[1]

Lars A. Carlson and Anders G. Olsson

King Gustaf V Research Institute and Department of Medicine,
Karolinska Hospital, Stockholm

Introduction

As hyperlipoproteinemia (HLP) may result in clinical diseases, serum-lipoprotein-lowering treatment has been introduced to diminish diseases or symptoms of diseases. The most utilized treatments are diet and drugs. Examples of diseases caused by HLP are disturbing xanthomas or recurring pancreatitis. As the relation between HLP and disease in these conditions is obvious, the indication of hypolipidemic treatment is clear-cut. Hypolipidemic drugs, however, are most often given with the intent to decrease atherosclerosis and its complications, e.g. acute myocardial infarction or intermittent claudication. Studies demonstrating a reduced morbidity or mortality in cardiovascular diseases by lipid-lowering drugs are scarce. Most studies concern secondary prevention, i.e. treatment after the appearance of overt disease. For example, we have recently showed [1] that administration of clofibrate 1 g twice daily plus nicotinic acid 1 g three times daily, which lowered serum cholesterol and triglycerides (TG) by 20 and 30%, respectively, significantly reduced the incidence of nonfatal myocardial infarctions while the mortality in ischemic heart disease has as yet not been affected. More studies are needed, especially in primary prevention, in order to establish the effect of therapeutic changes of serum lipoproteins on the development of atherosclerosis.

Thus the effect of a drug on lipoprotein concentrations is not a measurement of the ultimate clinical goal. In the future, direct demonstration of re-

---

[1] Supported by grants from the Swedish Medical Research Council (19X-204) and 'Petrus and Augusta Hedlunds Stiftelse'.

gression of human atherosclerosis by various hypolipidemic drugs will probably be mandatory. *Barndt et al.* [2] have recently demonstrated in a few patients with HLP that indeed it may be possible to reduce atherosclerosis.

Lipoproteins and Atherosclerosis

The various serum lipoproteins probably play different roles in the process of atherosclerosis. This is reflected in the serum lipoprotein concentrations as found in subjects with atherosclerosis. While increased concentrations of very-low-(VLDL) and low-(LDL) density lipoproteins are accompanied by increased atherosclerosis [i.e., increased incidence of acute myocardial infarction in a prospective study, 3, or higher prevalence of lipoprotein abnormalities in subjects with the disease, 4], low concentrations of high-density lipoproteins (HDL) are consistently observed in atherosclerosis [5, 6]. Logically, it may therefore be as important to increase the low HDL cholesterol concentrations as to decrease high levels of VLDL or LDL, in such patients.

Lipoprotein Interrelations

Serum lipoproteins are related to each other by several mechanisms. First, *statistic* relations exist between different lipoproteins in HLP. For example, there is a negative correlation between VLDL-TG and LDL cholesterol [7] and between VLDL-TG and HDL cholesterol [7]. There is no relation, however, between LDL cholesterol and HDL cholesterol. Second, *metabolically* the lipoproteins are also related. VLDL is considered to be degraded in the circulation to LDL_2 and there is evidence for a precursor-product relationship between the apolipoprotein B (apoB) of VLDL and LDL (p. 139). Exchange of apoprotein C (apoC) [8] and possibly cholesterol esters and TG may occur between VLDL and HDL. Treatment aiming at decreasing the concentration of one lipoprotein may therefore affect other lipoproteins in a wanted or unwanted direction. It is important to bear this in mind when studying the effect of drugs on serum lipids and lipoproteins. Third, *regulatory* relations between lipoproteins may exist. For example VLDL might stimulate the catabolism of LDL [9].

Against this background, it is necessary when evaluating the effect of a drug as a potential antiatherosclerotic agent through its action on serum

Table I. Effect of oxandrolone (7.5 mg/day) on serum lipoproteins (mmol/l) in HLP types II-B (n=4) and IV (n=16)

	Before	After	Change, %
Serum			
TG	4.12	2.87	−30**
Cholesterol	8.3	7.85	−2 n.s.
VLDL			
TG	2.97	1.93	−35**
Cholesterol	1.68	0.96	−43**
LDL			
TG	0.66	0.78	+18 n.s.
Cholesterol	4.64	5.46	+18*
HDL			
TG	0.28	0.18	−36**
Cholesterol	1.11	0.73	−35**

Significant difference: *$p<0.01$; **$p<0.001$. n.s. = Nonsignificant.

lipid concentrations, to study the effect on different serum lipoprotein concentrations and not to rely on total serum lipids.

An example of the necessity of lipoprotein analysis in the evaluation of drug effects is shown in table I. Oxandrolone which reduces serum total TG concentrations does not affect total serum cholesterol [10]. Determination of the effect on serum lipoproteins has demonstrated that oxandrolone increased LDL cholesterol and decreased HDL cholesterol, changes that are unwanted from the view of atherosclerosis prevention. Oxandrolone therefore can not be used as a lipid-lowering drug for atherosclerosis prevention, a conclusion that could not be reached when total serum TG and cholesterol concentrations were measured.

The serum concentration of a lipoprotein is determined by the rates of production into and elimination out of the blood stream. HLP can therefore be due to increased production, decreased removal of lipoproteins, or a combination of the two. The hypolipidemic action of a drug can be attributed to an effect on any of these mechanisms. Kinetic studies are required to answer the question of mode of action of a drug.

Combined dietary and drug treatment for reducing serum lipoproteins should be a lifelong treatment, as the aim is long-term prevention of atherosclerosis. This statement implies that the treatment should (1) effectively reduce atherogenic lipoprotein levels or increase antiatherogenic lipoproteins, (2) have few side-effects and (3) it should be well tolerated and convenient to use. As will follow below, available lipoprotein-lowering drugs do not fulfil these criteria today. Therefore, we are in urgent need of more effective and more convenient hypolipoproteinemic drugs with fewer side-effects.

Effects of Diet on Serum Lipoproteins

Elevated serum lipoprotein concentrations are generally treated by diet and/or drugs. Treatment of HLP often starts with a diet. The results are highly dependent on the effort put into the treatment. For example, if the treatment is conducted in a metabolic ward under meticulous control of food intake or on an outpatient basis. Also, distinction has to be made between short-term effects (weeks or months) and long-term effects of dietary advice. As an example, a study of the effect of dietary treatment in HLP is described. *Vessby and Lithell* [11] treated 120 patients with atherosclerotic cardiovascular disease on an outpatient basis for 2 months. 'Type-specific' diets were prescribed, with a calorie percent (average Swedish figures in parentheses) of fat of 35 (40–45), carbohydrate 45 (45–50) and protein 20 (11). In types II-A and B HLP, cholesterol intake was limited to less than 300 mg (normal diet 500–700 mg) and the polyunsaturated: saturated fatty acid ratio (P:S ratio) was kept at 2.0 (normally 0.2). In type IV HLP, the cholesterol intake was 300–500 mg and the P:S ratio 1.2. The results are shown in table II. Elevated VLDL concentrations (type II-B and IV HLP) decreased markedly on the diet, the fall of VLDL TG and cholesterol being approximately equal. Elevated LDL cholesterol also decreased significantly, so did LDL TG. Initially, normal LDL cholesterol concentrations did not change. The HDL fraction was not affected by diet with the exception of a decrease of HDL TG in type IV HLP.

Effects of Hypolipidemic Drugs on Serum Lipoproteins

Table III presents the general effects of three well-known drugs affecting serum lipoprotein concentrations – nicotinic acid, clofibrate (*p*-chlorophenoxyisobutyrate) and cholestyramine. Of these drugs, clofibrate is the easiest to take and has very few side-effects. In general, clofibrate is also the least

Table II. Effect of diet (percent) on serum lipoprotein concentrations in HLP [11]

	Type of HLP		
	II-A (n=21)	II-B (n=26)	IV (n=36)
Serum			
TG	− 7	−23**	−33***
Cholesterol	−13***	−13***	−11**
VLDL			
TG	− 1	−28**	−42***
Cholesterol	0	−34**	−43***
LDL			
TG	−14**	−12*	−13*
Cholesterol	−18***	−11**	− 5
HDL			
TG	0	− 8	−19**
Cholesterol	− 2	− 2	− 2

Significant difference: *p<0.05; **p<0.01; ***p<0.001.

Table III. General effects of nicotinic acid, clofibrate and cholestyramine on lipoprotein concentrations of different fractions in HLP

	VLDL triglycerides	LDL cholesterol	HDL cholesterol
Nicotinic acid	↓↓	↓↓[1]	↑[2]
Clofibrate	↓↓	↓[1]	↑[2]
Cholestyramine	(↑)	↓↓	(↓)

[1] When initially low in type IV HLP ↑, see text.
[2] When initially low.

effective. Clofibrate is usually prescribed at a dose of 1 g twice daily. Higher doses do not appear to give a better response [12]. We use clofibrate in patients with 'moderate' HLP and observe particularly good effects in type III HLP. As with most hypolipidemic regimes, the possibility of increased risk for gallstones [13], probably due to changed composition of the bile [14],

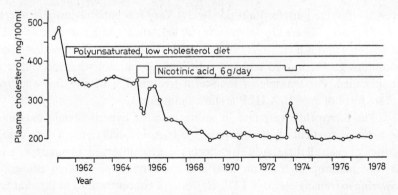

Fig. 1. Example of the effect of nicotinic acid (2 g t.i.d.) on the serum cholesterol concentration in a subject (case A.T., ♂, born 1911; myocardial infarction in 1961) with familial type II-A HLP.

should be considered. Furthermore, as with many hypolipididemic regimes, patients on clofibrate should be closely supervised for possible liver damage.

Cholestyramine should only be used when LDL is elevated and with some caution if VLDL is also elevated, as VLDL may be raised by bile acid sequestrants [15, 16]. We prescribe cholestyramine at a dose of 4 or 8 g twice daily and have found that the twice-daily dose is as effective as four times daily [17]. In addition, we administer vitamin K and folic acid. Cholestyramine often causes constipation, especially during the first 2–4 weeks of treatment. In about 25% of patients, this is sufficiently severe to discontinue the drug. Therefore, we always prescribe a bulk laxative together with cholestyramine when starting the treatment. The patient must be carefully instructed about the constipation and informed that the first weeks are the most troublesome. Thereafter, normal bowel habits may return. We treat patients with massive, often familial, type II-A HLP with cholestyramine.

Plain nicotinic acid is the most potent lipid-lowering drug. It is effective in type II-A, II-B, III, IV and V HLP and normalizes raised levels of chylomicrons, VLDL, LDL and 'β-VLDL' or late pre-β-lipoprotein (LPβ) [18] as well as decreased levels of HDL cholesterol. Nicotinic acid has a dose-response effect. The first maintenance dose is usually 1 g four times daily. Patients who tolerate this dose often tolerate higher doses. We have sometimes used 16–20 g of nicotinic acid in patients with severe HLP. Flush is an acute side-effect of nicotinic acid. As there is pronounced tachyphylaxis for the flush it would only very occasionally be a limitation for the treatment. However, the dosage must be built up slowly. Very few patients initially tolerate 1 g of the drug. There are several chronic side-effects with nicotinic acid. The most important is a rise of serum uric acid which eventually may cause gout. Monitoring of uric acid and institution of allopurinol when needed is recommended. An example of long-term treatment with nicotinic acid in a severe form of type II-A HLP is shown in figure 1.

The therapeutic response in severe cases of hyperlipidemia may vary. Sometimes, particularly with severe type II-A, one has to use a drug combination, or even all three. Additive effects are often observed. In our experience, the combination of nicotinic acid and cholestyramine is often efficient in lowering extremely elevated LDL cholesterol concentrations in familial type II-A HLP. However, gastrointestinal side-effects are very common with this combination.

Response and Pretreatment Serum Lipid Level

The response to diet, clofibrate and nicotinic acid treatment in HLP is related to pretreatment levels of total plasma lipids and in particular to lipoprotein lipids. This is exemplified in a study where the combined effect of diet and clofibrate (2 g daily) in 65 subjects with asymptomatic HLP of types II-A, II-B and IV [9] was determined. The average change in serum and lipoprotein lipids is shown in table IV. There were significant decreases in all fractions except HDL which rose by nearly 10%. Figures 2 and 3 show the relation between the decrease of serum cholesterol and TG levels and the pretreatment total serum cholesterol and TG concentrations. For cholesterol, the r value was about -0.7 and for TG close to -1. This indicates that only about 50% of the lowering of cholesterol was predicted by the pretreatment level. As will be shown below, the r values for cholesterol are considerably improved when related to lipoprotein cholesterol levels rather than total serum cholesterol.

For TG, there was nearly 100% prediction of the effect of treatment by the pretreatment level. This is most likely due to the fact that the TG is associated mainly with one lipoprotein class, VLDL.

There is an additional message in the data presented in figures 2 and 3 which we want to discuss. Neither regression line contains the origin when they are extrapolated to zero. This observation suggests that there is a lower limit at which treatment could reduce serum lipid levels. This theoretical limit was about 180 mg/100 ml of cholesterol and 1 mmol/l TG. From the

Table IV. Effect of clofibrate + diet on cholesterol (CH; mg/100 ml) and triglycerides (TG; mmol/l: mean values; n=65)

	Serum		VLDL		LDL		HDL	
	CH	TG	CH	TG	CH	TG	CH	TG
Pretreatment	307	3.9	56	2.8	195	0.8	49	0.3
Δ	−75	−2.3	−38	−2.1	−41	−0.2	+4	−0.1

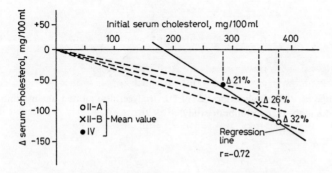

Fig. 2. Mean decrease of cholesterol (Δ serum cholesterol) plotted against initial serum cholesterol levels in a diet + clofibrate study. The regression line is obtained for individual values (n=65). The figure illustrates that the percentage decrease of serum cholesterol will decrease with decreasing initial level, as the regression line does not contain the origin.

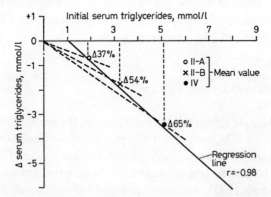

Fig. 3. Mean decrease of triglycerides (Δ serum triglycerides) plotted against initial serum triglycerides levels in a diet + clofibrate study. See legend of figure 3.

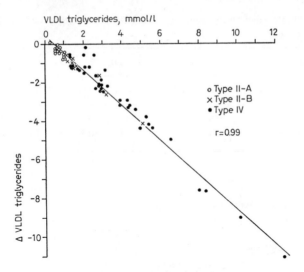

Fig.4. Change of VLDL tiglycerides (Δ VLDL triglycerides) plotted against initial VLDL triglycerides in the diet + clofibrate study.

practical point of view, there is one further aspect here. The fact that the regression lines do not contain the origin implies that the higher the pretreatment levels, the higher the percentage decrease. For cholesterol, the Δ% was 32% in type II-A and 21% in type IV. Considering the regression line, it would not be advisable to state that type II-A 'responded better' than type IV without mentioning that both Δs follow the same linear relation with the pretreatment level. The efficacy of a drug therefore, is best stated as a description of the line: its slope and intercept on the y-axis.

Response and Pretreatment Lipoprotein Level

The decrease of VLDL TG was predicted excellently by the pretreatment level (fig. 4). The decrease of LDL cholesterol was reasonably well predicted by the initial LDL cholesterol level (fig. 5). A most important aspect of the relation of the treatment effect to pretreatment LDL is that the regression line intersects the pretreatment LDL cholesterol at about 140 mg/100 ml. The implication of this observation is that when LDL cholesterol is below this value there occurs a rise in LDL cholesterol in response to treatment. Such

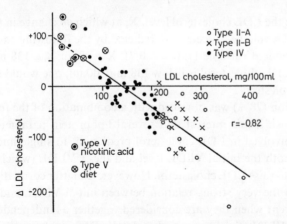

Fig.5. Change of LDL cholesterol (Δ LDL cholesterol) plotted against initial LDL cholesterol. The regression line intersects the x-axis at 138 mg/100 ml.

Table V. Regression analysis with the change (Δ) in LDL cholesterol in response to clofibrate + diet as dependent variable. Types II-A, II-B and IV HLP pooled (n=65)

	Independent variables												
	1	2	3	4	1	2	1	2	3	1	2	3	4
a	98	−88	−76	−8	52		30			12			
b	−0.71	17	−17	−0.20	−0.56	6.1	−0.52	32	29	−0.36	29	25	−0.75
t	11.5	7.8	6.1	8.8	6.9	2.7	6.6	3.2	2.7	3.9	3.0	2.4	3.1
R^2	0.65	0.49	0.41	0.54	0.71		0.74			0.78			

1 = LDL cholesterol; 2 = VLDL triglycerides: 3 = Δ VLDL triglycerides; 4 = LDL cholesterol: VLDL triglycerides.

rises in LDL cholesterol have been previously reported to be induced by diet or drugs [9, 11, 19, 20].

Multiple correlation analysis was performed on the data from the clofibrate + diet study to see which lipoprotein parameters could be related to the LDL response to treatment. The four parameters shown in table V demonstrated a significant regression coefficient with Δ LDL cholesterol as dependent variable. The initial LDL cholesterol had the highest predictability (R^2) followed by the ratio LDL:VLDL. The higher this ratio, the greater the fall in LDL. The usefulness of these data for single regression in table V is illustrated as follows: the LDL cholesterol level, X, at which *no* change in the LDL cholesterol level would be expected, is predicted by the following equation (table IV, independent variable 1): $0 = -0.71 \cdot X + 98$, or $X = 138$ mg/100 ml. Thus, when LDL cholesterol is below 138 mg/100 ml, one would expect a rise in LDL cholesterol.

The best prediction (78%) was obtained with a combination of the four variables. It is of considerable interest that several lipoprotein parameters were independently involved in LDL cholesterol prediction. In single linear regression analysis, both the initial VLDL level and the Δ VLDL (which is negative) would tend to raise LDL cholesterol. However, the latter correlation might only be due to the very strong relation between initial VLDL and Δ VLDL. This is apparent when they are considered together as independent factors. At any initial VLDL, the decrease induced in VLDL would contribute to a *lowering* of LDL cholesterol. As an example, in 2 patients, the levels of pretreatment LDL and VLDL are 100 mg/100 ml and 10 mmol/l, respectively, and the Δ VLDL are -4 and -8 mmol/l, respectively. The multiple-regression equation in table IV would predict the following changes in LDL cholesterol: patient 1: Δ LDL cholesterol $= (12 - 0.36 \cdot 100 + 29 \cdot 10 - 25 \cdot 4 - 0.75 \cdot 10) = +158.5$. Patient: 2 Δ LDL cholesterol $= (12 - 0.36 \cdot 100 + 29 \cdot 10 - 25 \cdot 8 - 0.75 \cdot 10) = +58.5$.

Apparently, the patient who had the greater fall in VLDL is predicted to show a considerably smaller rise in LDL cholesterol.

It is impossible to interpret these predictions at the metabolic level. It is apparent that a high VLDL level has a strong influence towards raising LDL cholesterol. The clinical benefit of lowering VLDL but raising LDL is, of course, not known. In addition, the rise in HDL which occurs must also be integrated. However, the prediction that irrespective of VLDL levels the rise in LDL is smaller when VLDL is maximally lowered, encourages aggressive treatment.

There was a small rise in HDL cholesterol. Linear-regression analysis showed that there is a correlation between Δ HDL cholesterol and pretreatment cholesterol ($r = -0.47$). The linear-regression analysis predicted that HDL cholesterol would rise if the initial level was 60 mg/100 ml or lower.

Comparison of the Lipid-Lowering Effects of Phenoxy Acids

Examples of the effects of some recently developed phenoxy acids on lipoprotein concentrations are give in figures 6 and 7. Studies on gemfibrozil

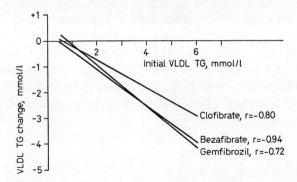

Fig. 6. Regression lines for the change of VLDL triglycerides against initial VLDL triglycerides after treatment with clofibrate (n = 29), bezafibrate (n = 29) and gemfibrozil (n = 26).

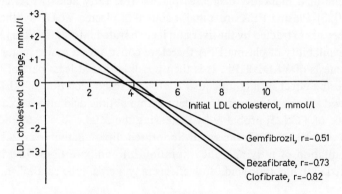

Fig. 7. Regression lines for the changes in LDL cholesterol against initial LDL cholesterol in the same studies as in figure 6.

[21] – 2.2-diurethyl-5(2.5-xylyloxy)-valeric acid – and bezafibrate [22] – 2-{4-[2-(4-chlorobenzamido)-ethyl]-phenoxy}-2-methyl-propionic acid – were performed on HLP subjects. The patients were advised at least 3 months prior to the study to maintain a diet with a high P:S ratio (above 1.5), low cholesterol (below 300 mg), and when overweight, to reduce the energy intake. If HLP persisted, drug treatment was started. The effect of bezafibrate (0.6 g daily, n = 29) was compared to clofibrate (1.5 g daily) [22]. The daily dose of gemfibrozil (n = 26) was 0.8 g. Bezafibrate decreased mean VLDL TG levels significantly more than did clofibrate (about 20%) and the effect of

gemfibrozil was of the same order of magnitude as bezafibrate (fig. 6). According to the regression analysis (fig. 7), the reciprocal effects on VLDL and LDL operate with all three drugs, the initial LDL concentration at which no change occurred being very similar. At higher initial LDL concentrations, bezafibrate and clofibrate had very similar effects. Using phenoxy acid derivatives, it seems to be difficult to find hypolipidemic drugs that are substantially more effective than clofibrate.

Effects of Drugs on VLDL Metabolism

Nicotinic acid and clofibrate effectively reduce increased VLDL TG concentrations; cholestyramine increase this lipoprotein, at least in type II-A HLP.

Nicotinic acid markedly decreases plasma free fatty acid (FFA) concentrations [23]. Plasma FFA are a major source of plasma VLDL synthesis in man as they are extracted by the liver and incorporated into VLDL TG [24]. Increased availability of plasma FFA therefore constitutes a major factor in the pathogenesis of VLDL HPL. It is then possible that at least part of the VLDL-lowering effect of nicotinic acid is via decreased secretion of VLDL into the blood stream. However, treatment with nicotinic acid also increases the clearance of TG [25] probably by increasing the catabolic rate of endogenous TG and VLDL [26]. Increased lipoprotein lipase activity [27, 28] as well as stimulation of fatty acid incorporation into adipose tissue [29] has been reported with nicotinic acid. Both effects may increase the catabolism of VLDL.

The effect of nicotinic acid on LDL kinetics was studied in 2 patients injected with labeled LDL before and after administration of nicotinic acid. The fractional catabolic rate of LDL remained unchanged [30]. We have observed similar results after a single injection of ^{125}I-LDL and institution of nicotinic acid during the turnover study (fig. 8). These results suggest that the decrease of plasma LDL during nicotinic acid treatment is due to decreased formation of these lipoproteins. It is possible that the inhibition of the mobilization of FFA from adipose tissue may decrease the formation of VLDL which, in turn, leads to decreased formation of LDL. In fact, treatment of patients with hypercholesterolemia causes first a decrease in VLDL followed by a decrease in LDL [31, 32].

The mode of action of clofibrate on VLDL metabolism is less clear. Different results have been reported regarding its effect on FFA levels [33, 34].

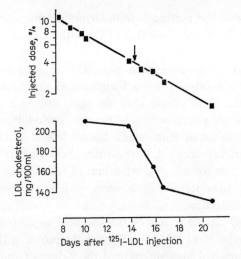

Fig. 8. Disappearance of radioactivity from plasma after i.v. injection of ^{125}I-labeled LDL (a) and cholesterol concentration in the LDL class (b) during nicotinic acid (4.5 g/day) treatment. Treatment was started at the arrow. Observe the fall in the concentration of LDL and unchanged fractional turnover indicating that nicotinic acid reduces the formation of LDL.

In isolated rat adipocytes, high concentrations of clofibrate inhibit fat-mobilizing lipolysis and antagonize the effects of noradrenaline or ACTH on cyclic AMP accumulation [35]. This study indicates that the mechanism behind the action of clofibrate on VLDL levels is a *decreased* rate of adipose tissue TG lipolysis due to lowering of the intracellular level of cyclic AMP. The decreased availability of FFA may cause a decrease of VLDL synthesis in the liver.

It has been reported very recently that one of the mechanisms by which clofibrate reduces VLDL is an *increase* of the activity of lipoprotein lipase in adipose tissue, thereby increasing the clearance of VLDL [36]. *Lithell et al.* [37] have reported 50% increases in skeletal muscle tissue lipoprotein lipase activity after clofibrate. These studies thus indicate that the effect of clofibrate on VLDL TG metabolism is via an increased removal rather than a decreased synthesis.

Similar to nicotinic acid, studies with ^{125}I-labeled LDL have shown that the fractional catabolism of apoB in LDL remains unchanged during treatment with clofibrate [38]. However, in this study, the LDL level remained unchanged while the apoB content of VLDL fell by approximately 50%. We therefore suggested that clofibrate had exerted its initial hypolipidemic effect by increasing the fractional catabolic rate of apoB in VLDL without changing the rate of synthesis [38].

It is interesting that many compounds, such as insulin and catecholamines, exert opposite effects on fat-mobilizing lipolysis and lipoprotein lipase activity in adipose tissue. Both effects have the same influence on VLDL concentration, although one affects removal (stimulated or inhibited) and the other production (inhibited or stimulated). Insulin lowers VLDL concentration while catecholamines increase it. With nicotinic acid, clofibrate, and clofibrate analogues [i.e. gemfibrozil, 39], inhibition of fat-mobilizing lipolysis in adipose tissue and stimulation of lipoprotein lipase is observed. Both would lower VLDL concentration.

In a metabolic ward study on the effects of the bile acid sequestrant agent, cholestyramine, we noted that VLDL TG increased *abruptly* after the drug was administered and remained high throughout the 9 days of study [16]. Increased VLDL levels during treatment with bile-acid-sequestering drags has previously been observed [15]. VLDL cholesterol also increased but to a lesser degree resulting in decreased ratio of cholesterol to TG in VLDL. This observation implies that after cholestyramine the VLDL particles were larger; the ratio of cholesterol to TG may reflect the proportion of surface to core constituents of the particles (see p. 139). Indeed, *Witztum et al.* [40] have found larger TG-rich particles after colestipol, as determined by column chromatography. The effect of bile-acid-sequestering drugs (cholestyramine) and the ensuing stimulation of bile acid synthesis [41] on plasma endogenous TG kinetics was studied by *Angelin et al.* [42]. All subjects with type II HLP treated with cholestyramine displayed increased biosynthesis and elevated fractional turnover rates of TG. No consistent effect of the drug was encountered in type IV HLP. Bile acid sequestration thus stimulated the production of plasma TG – and presumably VLDL TG. Also, interruption of the bile acid enterohepatic circulation has been shown to increase the synthesis of cholesterol in the liver. Thus the increased catabolism of LDL cholesterol by bile-acid sequestering drugs [43] seems to be compensated for by increased VLDL synthesis, at least in part.

Effects of Drugs on LDL Metabolism

In type II-A HLP, the bile acid sequestrant agent, cholestyramine, is a most effective LDL-cholesterol-lowering drug in clinical practice. With

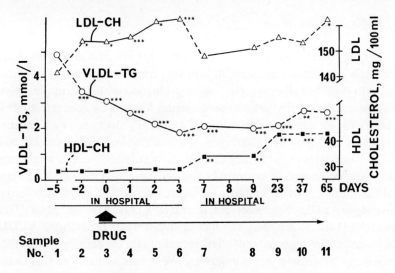

Fig. 9. Effect of drug treatment on LDL cholesterol (CH), VLDL triglycerides and HDL cholesterol in 30 subjects with type IV HLP.

nicotinic acid, similar effects can be obtained only if high dosages are used. These drugs may reduce LDL cholesterol by 25–30%.

Studies with labeled LDL have shown that cholestyramine acts by increasing the fractional catabolism of LDL [44] while nicotinic acid lowers LDL by reducing its rate of formation from VLDL (see above on VLDL). It is important to realize, however, that only very few studies were conducted, and were moreover restricted to patients with type II-A HLP.

To clarify the mechanism by which LDL increases in patients with type IV HLP (see above), we have studied the effect of three drugs – clofibrate, 1.5 g daily, nicotinic acid, 4.5 g daily, and clofibrate, 1.5 g + pyridylcarbinol 0.075 g daily – on serum lipoprotein concentrations in 30 subjects with type IV HLP [45] while on a metabolic ward (fig. 9). No differences in the effects on lipoprotein concentration were found. Consequently, the results were pooled. The design of the study and the effects are shown in figure 9. The decrease of VLDL and increase of LDL levels occurred simultaneously, indicating a close metabolic relationship between the two. One possible explanation for this phenomenon is increased transfer of apoB from VLDL to LDL. This suggestion is consistent with the findings of an increased VLDL-B clearance, as demonstrated in a kinetic study with labeled LDL B protein [38].

Practical Implications

The fact that 'hypolipidemic' drugs often have diverse effects on serum lipoprotein concentrations, some of potential benefit and some of potential harm with regard to atherogenesis, makes evaluation of these drugs complex. It can be concluded that estimation of serum lipoprotein concentrations is necessary for an adequate evaluation of a drug. By doing so, we have concluded that oxandrolone probably is of no benefit because of the increase of LDL cholesterol and decrease of HDL cholesterol. However, even drugs such as clofibrate and its derivatives and nicotinic acid may under certain conditions, exert unwanted effects on lipoprotein concentrations. Cholestyramine lowers LDL cholesterol but increases VLDL TG, decreases HDL cholesterol [16] and increases the flux of lipoproteins through the VLDL-LDL cascade. The importance of all these findings for the development and/or regression of atherosclerosis is unknown. We conclude therefore that in the future, methods for the quantitation of the effects of a drug, not only on serum lipoproteins, but also directly on the amount of atherosclerosis will be needed.

References

1 Carlson, L.A.; Danielson, M.; Ekberg, I.; Klintemar, B., and Rosenhamer, G.: Reduction of myocardial reinfarction by the combined treatment with clofibrate and nicotinic acid. Atherosclerosis *28:* 81–86 (1977).
2 Barndt, R., jr.; Blankenhorn, D.H.; Crawford, D.W., and Brooks, S.H.: Regression and progression of early femoral atherosclerosis in treated hyperlipoproteinemic patients. Ann. int. Med. *86:* 139–146 (1977).
3 Carlson, L.A. and Böttiger, L.E.: Ischaemic heart-disease in relation to fasting values of plasma triglycerides and cholesterol. Lancet *i:* 865–868 (1972).
4 Carlson, L.A. and Ericsson, M.: Quantitative and qualitative serum lipoprotein analysis. 2. Studies in male survivors of myocardial infarction. Atherosclerosis *21:* 435–450 (1975).
5 Miller, N.E. and Miller, G.J.: High density lipoprotein and atherosclerosis. Lancet *i:* 1033 (1975).
6 Gordon, T.; Castelli, W.P.; Hjortland, M.C.; Kannel, W.B., and Dawber, T.R.: High density lipoprotein as a protective factor against coronary heart disease. The Framingham Study. Am. J. Med. *62:* 707–714 (1977).
7 Olsson, A.G. and Carlson, L.A.: Studies in asymptomatic primary hyperlipidaemia. I. Types of hyperlipoproteinaemias and serum lipoprotein concentrations, compositions and interrelations. Acta med. scand., suppl. 580 (1975).
8 Eisenberg, S.; Bilheimer, D.W., and Levy, R.I.: The metabolism of very low density lipoprotein proteins. II. Studies on the transfer of apoproteins between plasma lipoproteins. Biochim. biophys. acta *280:* 94 (1972).

9   Carlson, L.A.; Olsson, A.G.; Orö, L.; Rössner, S., and Walldius, G.: Effects of hypolipidaemic regimes on serum lipoproteins; in Schettler and Weizel, Atherosclerosis III. Proc. 3rd Int. Symp., p. 768 (Springer, Berlin 1974).

10 Olsson, A.G.; Orö, L., and Rössner, S.: Effects of oxandrolone on plasma lipoproteins and the intravenous fat tolerance in man Atherosclerosis 19: 337–346 (1974).

11 Vessby, B. and Lithell, H.: Dietary effects on lipoprotein levels in hyperlipoproteinemia. Delineation of two subgroups on endogenous hypertriglyceridemia. Artery 1: 63–85 (1974).

12 Olsson, A.G.; Orö, L., and Rössner, S.: Dose-response effect of single and combined clofibrate (Atromidin®) and niceritrol (Perycit®) treatment on serum lipids and lipoproteins in type II hyperlipoproteinaemia. Atherosclerosis 22: 91–101 (1975).

13 The Coronary Drug Project: Clofibrate and niacin in coronary heart disease. J. Am. med. Ass. 231: 360–381 (1975).

14 Pertsemlidis, D.; Panveliwalla, D., and Ahrens, E.H., jr.: Effects of oral contraceptives on the gallbladder bile of normal women. New Engl. J. Med. 294: 189–192 (1976).

15 Miller, N.E. and Nestel, P.J.: Differences among hyperlipoproteinaemic subjects in the response of lipoprotein lipids to resin therapy. Eur. J. clin. Invest. 5: 241–247 (1975).

16 Olsson, A.G. and Dairou, F.: Acute effects of cholestyramine on serum lipoprotein concentrations in type II hyperlipoproteinaemia. Atherosclerosis 29: 53–61 (1978).

17 Olsson, A.G.: The effect of cholestyramine (Questran), 8 g twice-daily, on serum lipoprotein concentrations in type II hyperlipoproteinaemia. Pharmatherapeutica 1: 583–587 (1977).

18 Carlson, L.A.: The effect of nicotinic acid treatment on the chemical composition of plasma lipoprotein classes in man; in Holmes, Carlson and Paoletti, Drugs affecting lipid metabolism; vol. 4, p. 327 (Plenum Press, New York 1969).

19 Strisower, E.H.; Adamson, G., and Strisower, B.: Treatment of hyperlipidemias. Am. J. Med. 45: 488–501 (1968).

20 Wilson, D.E. and Lees, R.S.: Metabolic relationships among the plasma lipoproteins. Reciprocal changes in the concentrations of very low density lipoproteins in man. J. clin. Invest. 51: 10151–1057 (1972).

21 Olsson, A.G.; Rössner, S.; Walldius, G., and Carlson, L.A.: Effect of gemfibrozil on lipoprotein concentrations in different types of hyperlipoproteinaemia. Proc. R. Soc. Med. 69: 28–31 (1976).

22 Olsson, A.G.; Rössner, S.; Walldius, G.; Carlson, L.A., and Lang, P.D.: Effect of BM 15.075 on lipoprotein concentrations in different types of hyperlipoproteinaemia. Atherosclerosis 27: 279–287 (1977).

23 Carlson, L.A. and Orö, L.: The effect of nicotinic acid on the plasma free fatty acids. Demonstration of a metabolic type of sympathicolysis. Acta med. scand. 172: 641–645 (1962).

24 Robinson, D.S.: The function of plasma triglycerides in fatty acid transport; in Florkin and Stotz, Comprehensive biochemistry. Lipid metabolism, chapter 1, p. 51 (Elsevier, Amsterdam 1970).

25 Boberg, J.; Carlson, L.A.; Fröberg, S.; Olsson, A.G.; Orö, L., and Rössner, S.: Effects of chronic treatment with nicotinic acid on intravenous fat tolerance and postheparin lipoprotein lipase activity in man; Gey and Carlson, Metabolic effects of nicotinic acid and its derivatives, p. 465 (Huber, Bern 1971).

26 Rössner, S.; Boberg, J.; Carlson, L.A.; Freyschuss, U., and Lassers, B.W.: Comparison between fractional turnover rate of endogenous plasma triglycerides and of Intralipid ® (intravenous fat tolerance test) in man. Eur. J. clin. Invest 4: 1–6 (1974).

27 Nikkilä, E.A.: Effect of nicotinic acid on adipose tissue lipoprotein lipase and removal rate of plasma triglycerides; in Gey and Carlson, Metabolic effects of nicotinic acid and its derivatives, pp. 487–496 (Huber, Bern 1971).

28 Otway, S.; Robinson, D.S.; Rogers, M.P., and Wing, D.R.: The effects of nicotinic acid and of glucose on adipose tissue and heart clearing factor lipase activities and free fatty acid concentrations in the starved rat; in Gey and Carlson, Metabolic effects of nicotinic acid and its derivatives, pp. 497–514 (Huber, Bern 1971).

29 Walldius, G.: Fatty acid incorporation into human adipose tissue (FIAT) in hypertriglyceridaemia. Methodological, clinical and experimental studies. Acta med. scand., suppl. 591: (1976).

30 Langer, T. and Levy, R.I.: The effect of nicotinic acid on the turnover of low density lipoproteins in type II hyperlipoproteinemia; in Gey and Carlson, Metabolic effects of nicotinic acid and its derivatives, pp. 641–647 (Huber, Bern 1971).

31 Carlson, L.A.; Orö, L., and Östman, J.: Effect of a single dose of nicotinic acid on plasma lipids in patients with hyperlipoproteinaemia. Acta med. scand. 183: 457–465 (1968).

32 Carlson, L.A.; Orö, L., and Östman, J.: Effect of nicotinic acid on plasma lipids in patients with hyperlipoproteinaemia during the first week of treatment. J. Atheroscler. Res. 8: 667–677 (1968).

33 Rifkind, B.M.: Effect of CPIB ester on plasma free fatty acid levels in man. Metabolism 15: 673–675 (1966).

34 Duncan, C.H.; Best, M.M., and Robertson, G.L.: A comparison of the effects of ethylchlorophenoxyisobutyrate and nicotinic acid on plasma free fatty acids. Lancet i: 191–193 (1965).

35 Carlson, L.A.; Walldius, G., and Butcher, R.W.: Effect of chlorophenoxyisobutyric acid (CPIB) on fat-mobilizing lipolysis and cyclic AMP levels in rat epididymal fat. Atherosclerosis 16: 349–357 (1972).

36 Taylor, K.G.; Holdsworth, G., and Galton, D.J.: Clofibrate increases lipoprotein-lipase activity in adipose tissue of hypertriglyceridaemic patients. Lancet ii: 1106–1107 (1977).

37 Lithell, H.; Boberg, J.; Hellsing, K.; Lundqvist, G., and Vessby, B.: Increase of lipoprotein lipase activity in human skeletal muscle tissue after clofibrate treatment. Abstract. 6th Int. Symp. on Drugs Affecting Lipid Metabolism, Philadelphia 1977.

38 Ballantyne, F.C.; Ballantyne, D.; Olsson, A.G.; Rössner, S., and Carlson, L.A.: The effect of clofibrate therapy on low density lipoprotein metabolism in type IV hyperlipoproteinaemia. Pharmatherapeutica 1: 481–492 (1977).

39 Carlson, L.A.: Effect of gemfibrozil in vitro on fat-mobilizing lipolysis in human adipose tissue. Proc. R. Soc. Med. 69: 101–103 (1976).

40 Witztum, J.L.; Schonfeld, G., and Weidman, S.W.: The effects of colestipol on the metabolism of very-low-density lipoproteins in man. J. Lab. clin. Med. 88: 1008–1018 (1976).

41 Einarsson, K.; Hellström, K., and Kallner, M.: The effect of cholestyramine on the elimination of cholesterol as bile acids in patients with hyperlipoproteinaemia type II and IV. Eur. J. clin. Invest. 4: 405–410 (1974).

42 Angelin, B.; Einarsson, K.; Hellström, K., and Leijd, B.: Bile acid kinetics in relation to endogenous triglyceride formation in various types of hyperlipoproteinemia. Acta med. scand., suppl. 610 (1977).
43 Miller, N.E.; Clifton-Bligh, P., and Nestel, P.J.: Effects of colestipol, a new bile-acid-sequestring resin, on cholesterol metabolism. J. Lab. clin. Med. *82:* 876–890 (1973).
44 Levy, R.I. and Langer, T.: Hypolipidemic drugs and lipoprotein metabolism; in Holmes, Paoletti and Kritchevsky, Pharmacological control of lipid metabolism, p. 155 (Plenum Publishing, New York 1973).
45 Carlson, L.A.; Olsson, A.G., and Ballantyne, D.: On the rise in low density and high density lipoproteins in response to the treatment of hypertriglyceridaemia in type IV and type V hyperlipoproteinaemias. Atherosclerosis *26:* 603–609 (1977).

L.A. Carlson, MD, King Gustaf V Research Institute and Department of Medicine, Karolinska Hospital, S–104 01 Stockholm 60 (Sweden)

Subject Index

Abetalipoproteinemia 170, 186
Apolipoprotein A-I 203
  in HDL 201, 203
    kinetic analysis 97–99, 208
  LCAT activator 44
  in LCAT deficiency 45, 46
  origin, intestine 207
    liver 206, 207
  plasma levels 204
  purification 203
Apolipoprotein A-II 203
  in HDL 201, 203
    kinetic analysis 97–99, 208
  in LCAT deficiency 45, 46
  plasma levels 204
Apolipoprotein B
  binding to heparin 116
  in IDL 153–156
    kinetic analysis 91, 92
  in LCAT deficiency 46
  in LDL 147, 148, 168–170
    kinetic analysis 92–95
  metabolism 99, 100
    in hypertriglyceridemia 157, 158, 250, 251
  molecular weight 168
  receptors 56, 59, 153–156, 168, 173, 184, 222–229
  in VLDL 140, 141
    during lipolysis 147, 148
    kinetic analysis 83–88

Apolipoprotein C-I
  in HDL 201, 204
  in LCAT deficiency 45, 46, 55
Apolipoprotein C-II
  activation of lipoprotein lipase 20, 112–114, 129, 153, 155
  deficiency, familial 160
  exchange between lipoproteins 146, 147
  in HDL 201, 204
  in LCAT deficiency 45, 46
  in VLDL 141
    during lipolysis 146, 147
Apolipoprotein C-III
  exchange between lipoproteins 141–147
  in HDL 201, 204
  in LCAT deficiency 46, 55
  in VLDL 141
    during lipolysis 146, 147
Apolipoprotein D 201, 204
Apolipoprotein E
  cholesterol ester transfer 54
  in HDL 201, 204
    transfer to VLDL 54
  heparin binding to 116
  in LCAT deficiency 45, 46
  receptor 59, 231
  in remnants 154–156, 218, 219
  in VLDL 141, 148
Apolipoprotein F 204
Apolipoproteins, kinetic analysis 67–108

Subject Index

Arachidonic acid 43, 56–59
  in rats plasma transport 56–59
Arterial smooth muscle cells 187–191, 224–228
Atherogenesis 187–191, 224–228

Carbohydrate feeding 174, 208
Cholesterol
  exchange 149, 150
  in HDL 205
  reverse transport 55, 56, 179–181
  in VLDL 149, 150
Cholesteryl ester exchange protein 47–50
Cholesteryl ester hydrolase 11, 224–228
Cholestyramine 241–243, 252, 253
Chylomicrons 109–138
  composition 109
  fusion with cells 121–129
  and lipoprotein lipase 111, 114–116
  remnants 30, 31, 52, 216–218, 220, 222
  size 109
  triglyceride hydrolase, hepatic 17
Clofibrate 241–254
  and lipoprotein lipase 30, 250, 251

Erythrocytes 47, 48

Familial hypercholesterolemia (FH) 171–173, 178, 186, 187, 189, 190, 223, 252, 253
Fatty acids transfer 116–129
  inactivation of LPL 129–131
Fluorescence lipids 117–119

Heparan sulfate 115
High density lipoproteins (HDL)
  apoA-I, kinetic analysis 97–99
  apoC, kinetic analysis 100, 101
  apoproteins 201
  catabolism 208–210, 229–231
  cholesterol, in hyperlipoproteinemia 206
    plasma levels 205, 206
    reverse transport 55, 56, 179, 180
  cholesteryl esters 49, 50
  composition 200, 201
  fusion 119, 120

and LCAT 43–45, 49, 50, 207
in LCAT deficiency 44, 45, 152, 207
metabolism 200–215
origin, from intestine 206, 207
  from liver 207
  from VLDL 152, 153
receptors 56, 58, 229–231
subfractions 200, 201
and VLDL, effect of lipolysis 120, 152, 153
High density lipoprotein C (HDL$_c$) 167, 180, 202, 231
Hyperlipoproteinemia 20, 29, 158, 160, 174, 205, 206
  treatment 238–257
Hyperlypoproteinemia, type I 20, 29, 160
  type II, see Familial hypercholesterolemia
  type III 29, 158, 205, 242, 243
  type IV 159, 160
    treatment 241, 243, 244, 247–252
  type V 29, 159, 160, 205
Hypobetalipoproteinemia 186

Intermediate density lipoproteins (IDL) 91, 92, 143–145, 157–159, 170–175, 218, 220, 239, 240

LCAT deficiency, familial 29, 44–48, 152, 207
Lecithin:cholesterol acyltransferase (LCAT) 41–46, 153, 206, 207
  and apoprotein transfer 45–47
  and cholesteryl ester transfer 49, 50
  and HDL 43–45, 153, 206, 207
  and lipoprotein lipase 51, 52, 153
  and plasma lipids 42
  and VLDL and chylomicrons 51–54
Lipase, liver cytosol 9, 10
  liver lysosomes 10–13
  liver microsomes 7–9
  liver mitochondria 6, 7
  liver plasma membranes 13–16
Lipoprotein lipase
  activation by apoC-II 20, 112–114
  activation by apoC-II fragments 112, 113

Subject Index

and chylomicrons phospholipid hydrolysis 111, 112
and chylomicrons triglyceride hydrolysis 111
and clofibrate 250
and endothelial cells 115, 116
and estrogens 26
in heart cells 217, 218
in heart perfusion 146–153
heparin releasable 19–27
and heparin-sepharose 22, 23, 115
in hypertriglyceridemia 28, 29
in activation by fatty acids 129–131
and LCAT 51
in liver diseases 28
and nicotinic acid 250
phospholipase activity 148, 149
plasma levels 26, 27
in renal diseases 28
tissue origin 20
and VLDL and chylomicron metabolism 217, 218

Lipoproteins
catabolism 216–237
fusion 119–121
kinetics 67–108
    cascade analysis 79–81
    compartmental analysis 72–74
    computation models 71, 72
    residence time 75, 76

Low density lipoprotein (LDL) 166–199
apoB, kinetics 92–95
    turnover rate 178
and atherosclerosis 187–191, 224–228
catabolism 222–229
cationized 190, 223–226
composition 168, 169
degradation in liver 181–183
degradation in tissues 183–185, 228, 229
extravascular pool 176, 177
heterogeneity 166–168
in LCAT deficiency 45–47
in lymph 178, 181
origin from liver 171–175
origin from VLDL 93, 143–145, 157–159, 170, 171, 228, 239, 241

receptor, *see* Apolipoprotein B receptor
size 168
structure 145, 168
triglycerides 95–97
LP (a) 202
Lysolecithin, origin during lipolysis 119–121, 148, 149
Lysophospholipase
liver lysosomes 10
liver mitochondria 6, 7
liver plasma membrane 14

Monoacylglycerol 116–129
Monoglyceride lipase
heparin releasable 18
liver lysosomes 10
liver microsomes 7, 8
liver perfusate 18

Nicotinic acid 208, 241–243, 250

Oxandrolone 30, 240

Phosphatidylcholine (lecithin)
exchange, in HDL 118
    between VLDL and HDL 146, 148, 149
and LCAT 42
Phospholipase
liver lysosomes 10
liver microsomes 8, 9
liver mitochondria 6, 7
liver plasma membranes 14
plasma (heparin releasable) 20, 21
Phospholipid exchange proteins 121, 148
Post heparin lipolytic activity (PHLA) 5, 19–27
enzymes 21, 22
in liver diseases 28

Sialic acids (in LDL) 169
Tangier disease 152, 201, 205, 208
Tetramethylurea 140
Triglyceride lipase (hepatic) 5–40
and bezafibrate 30
and clofibrate 30
in diabetes 28

and estrogens 29
in humans 27
in hyperlipoproteinemia 28, 29
inhibition by HDL 17
inhibition by plasma 143, 144
and LCAT deficiency 29
in liver cytosol 10
in liver diseases 28
in liver lysosomes 10
in liver plasma membranes 14–16
in liver perfusate 17–19
and oxandrolone 30
in plasma 21, 23–27
purification 25–27
and renal diseases 28, 29
and VLDL 143, 144
Triglyceride transport
in chylomicrons 109–138
in VLDL 142–145

Very low density lipoprotein (VLDL) 139–165
apoB, content 140–142, 146, 147, 153, 155
kinetics 83–88, 157, 158
synthetic rate 157, 158
apoC, content 140–142, 146, 147, 153, 155
kinetics 100, 101
synthetic rate 158
apoE, content 140–142, 148
turnover rate (rats) 159
catabolism 155, 218, 220, 222
cholesteryl esters 43, 145
composition 140–142
conversion to LDL 83–92
and drugs 250–252
free cholesterol in 149, 150
and HDL 152, 153
interaction with cells 155
interaction with LPL 142, 143
interaction with triglyceride lipase (hepatic) 17
and LCAT 43–45
in LCAT deficiency 45–47
origin 139, 173
phospholipids 148, 149
remnants 50, 158, 159, 172, 216–220
size 139
structure 141
subfractions 140–142, 153–155
triglyceride transport 88–91, 142–145
turnover 79–83

Wolman's disease 12, 223

173387